BASIC WATER
TREATMENT

BASIC WATER TREATMENT

THIRD EDITION

Chris Binnie
Martin Kimber
George Smethurst

Thomas Telford

Published by Thomas Telford Publishing, Thomas Telford Ltd,
1 Heron Quay, London E14 4JD.
URL: http://www.thomastelford.com

Co-published by IWA Publishing, Alliance House, 12 Caxton Street,
London SW1H 0QS, UK
Telephone: +44 (0) 20 7654 5500; Fax: +44 (0) 20 7654 5555;
Email: publications@iwap.co.uk
Web: www.iwapublishing.com

Distributors for Thomas Telford books are
USA: ASCE Press, 1801 Alexander Bell Drive, Reston, VA 20191-4400,
USA
Japan: Maruzen Co. Ltd, Book Department, 3–10 Nihonbashi 2-chome,
Chuo-ku, Tokyo 103
Australia: DA Books and Journals, 648 Whitehorse Road, Mitcham 3132,
Victoria

First published 1979
Second edition 1988
Reprinted with amendment 1990
Reprinted 1992, 1996, 1997, 1998, 2001
This edition 2002

A catalogue record for this book is available from the British Library
ISBN: 0 7277 3032 0

Typeset by Gray Publishing, Tunbridge Wells, Kent
Printed and bound in Great Britain by MPG Books, Bodmin, Cornwall

Contents

Foreword

I am delighted to see this timely third edition of *Basic Water Treatment*. It is far more than just an update of the earlier editions, as it includes processes, such as GAC filtration and ozonization, which at the time of first edition in 1979 would have been considered advanced, and which are now commonplace in the developed world. The chapter on membranes is welcome as they are becoming increasingly attractive, particularly for small installations.

The requirements for improved water treatment throughout the world is being driven by demanding numerical drinking water quality standards, based on both health and aesthetic considerations, the foundation for which are the World Health Organization Guidelines. In the UK, and in some other countries, this is accompanied by close scrutiny by Inspectorates, to ensure that those standards are met. These influences require not only that the design of treatment is right, but that the plant is operated effectively all of the time. This in turn requires good design engineers and knowledgeable operatives, who need good education and training material, to which this book can make a significant contribution.

The extensive use of GAC and ozone in the UK has been driven largely by the requirement to meet the pesticide standard of 0.1 µg/l. Many have argued that this was an unnecessary expense, but today it is becoming accepted as the means of achieving consumer confidence in tap water. Consumers are not generally impressed by the low risk arguments, they don't like the thought of pesticides being present in their drinking water, and the introduction of these processes is giving protection against other organics, such as residues from pharmaceutical drugs.

But we must not forget the value and importance of what I call 'conventional water treatment', namely coagulation, separation and filtration. The importance of effective and reliable particle removal processes has been reinforced by the threat from *Cryptosporidium*. Experience has shown that conventional treatment, when operated reliably, is effective and, for that reason, the regulations in England and Wales set a treatment standard, not a health standard. This is another example of public confidence relying on effective water treatment.

The above examples have merely touched on the importance of water treatment and thus the value of this book. I am sure that students and practitioners alike will benefit from it.

Michael Rouse CBE
Chief Inspector of The Drinking Water Inspectorate

Preface to the third edition

The third edition of this book, which has become one of the accepted primers on water treatment, has involved greatly expanding the processes covered and significant revisions to the coverage of the 'traditional' processes. This has been driven by the enormous changes that have occurred since the publication of the first edition. This is a relatively slim volume covering water treatment, a very broad subject. Thus a degree of subjectivity has had to be exercised over what should be covered. The third edition has trimmed the coverage of some of the older forms of clarifiers and filters, that are now rarely encountered, and has much additional coverage of newer processes.

Basic Water Treatment is aimed at university students, at practising water treatment engineers, for whom it will be a useful reference book, and at mechanical engineers and chemists who need to put their specialized knowledge into a broader context; it will also be of interest to those managers and non-technical staff who wish to understand some of the quality and technical issues relating to water treatment. It is not a prescriptive handbook on water treatment plant design; rather it provides essential background and is a useful first choice reference book for many aspects of water quality and treatment.

It is important that all involved in water quality and water treatment do not adopt too blinkered an approach. In Western Europe we now have to focus very closely on compliance with prescriptive water quality standards, although the Water Framework Directive and public concerns over the environment require a wider consideration of water resources and water demands. However, over much of the world, water engineers should consider, or at least be aware of, much wider issues, weighing the costs of high quality water against the overriding need to supply safe water to as many as possible. *Basic Water Treatment* primarily relates to water treatment in Western Europe and North America but it also takes account of treatment in other developed areas, and in developing and less developed areas.

1: Introduction

Water-supply engineering is a large and growing field of engineering. Up until the second half of the twentieth century water supply was rather a simple field. Water was abstracted, preferably from a reservoir with a protected catchment, or from wells, and after generally minimal treatment was put into the distribution network. Some supplies were derived from rivers and these supplies required more treatment. However where relatively poor-quality lowland rivers were used as a source, it was common to provide extensive storage prior to treatment, to allow self-purification of poor-quality water, and the subsequent treatment was often slow sand filtration, a process simple to operate and robust.

Distribution systems were also simple. In the UK we generally pumped water to water towers or service reservoirs located at high points, and from there the water served the supply zones by gravity. There was little or no flow measurement, except at treatment works, and leakage was generally not a major concern, not least because leakage rates could not be determined. However as far back as 1885, Alexander Binnie, forefather of one of the authors, was working to introduce metered water-supply zones with measurement and recording of night flows in Bradford, to allow wastage and leakage to be reduced.[1] Bradford used inspectors to visit houses to check water pipes and fittings.

The water-quality requirement was that the water be wholesome. In practice, the only quality standards were microbiological and regular water sampling and analyses were limited to microbiological examination and testing of chlorine levels. Universal chlorination of public drinking water was not introduced in the UK until after the Croydon typhoid outbreak of 1936.

The major concern during the first half of the twentieth century was to meet the ever-increasing demand for water. This was driven both by a steady increase in population served and by increasing per capita demands as consumers became ever more affluent and acquired indoor toilets, baths, washing-machines, cars, and garden hoses. Water supply was a public service and was seen as a basic right. Public funding was available and was used to construct the infrastructure to meet projected demands. Water supplies were generally unmetered and there was no economic incentive to reduce water use; environmental concerns had not developed; and there was little public objection to the construction of additional infrastructure. The infrastructure was often of the highest quality, and much of the works constructed between the late nineteenth century and the early twentieth century continues to serve water consumers today. A similar situation prevailed over much of Europe and other developed countries.

There were enormous changes over the period 1975–2000. In the UK some of the factors responsible include:

- the adoption of level of service indicators for public water supply—notably water-quality standards and standards for pressure of water supply;
- the rationalization of much of the water-supply industry into river catchment-based groupings;
- the introduction of numerical standards for water quality, derived from the 1980 European Drinking Water Directive;
- the privatization of the water industry and the introduction of the associated regulatory structure;
- the widespread introduction of water metering, now universal for industrial and commercial supplies and used for a significant proportion of domestic supplies in England and Wales;
- increasing environmental awareness among the general public leading to extreme difficulties in promotion of major infrastructure works and in increasing the quantity of groundwater used in water supply; and
- two major droughts affecting large areas of the country.

The consequence has been a near total transformation of the UK water-supply industry, and at the time of writing this transformation is still underway. Most other developed countries have had or are having a similar transformation in their water-supply industries.

What effect have these changes had on water engineers? Arguably the most important change has been the need to adopt a holistic approach to water supply. It is no longer possible or acceptable to consider the various components making up a water supply in isolation. The need for environmental protection and the difficulties of constructing new dams has meant that it is no longer possible to meet the water demands that would arise from an unfettered growth in demand. Thus means have to be adopted to reduce demands and to lower the amount of water lost through leaks. The adoption of extensive new physical and chemical water-quality standards which apply at the point of delivery to the consumer has meant not only additional water treatment, but the close examination of water-distribution systems to ensure that the water entering the distribution system does not deteriorate unacceptably as it travels to the point of use. Environmental considerations also impact on treatment processes used, with the problems and costs associated with the acceptable treatment and disposal of wastes arising from treatment becoming ever more significant.

Water supply in the developing world reflects some of the above, but with other problems derived from the particular financial, social, and institutional constraints applying to each country. In particular there are often very high rates of demand growth associated with increasing urbanization, high rates of population growth, and increased wealth leading to increased ownership of water-using appliances. Common problems include: very high per capita water production, a shortage of economic water resources; and low water prices leading to waste and a lack of money for operating and maintaining systems.

In much of Europe there may be complaints about the cost of water, but it is generally accepted that it is essential to have the highest-quality water, almost regardless of cost. In poorer parts of the world such an approach is not possible and standards and the costs of meeting standards may have to be looked at in the context of limited resources and comparative risks. The World Health Organization (WHO) has set guideline standards for carcinogens on the basis of one additional cancer case per 100 000 population receiving water at the guideline value over 70 years. In countries with limited resources and short life expectancies it may be difficult to justify investing heavily in expensive water treatment in preference to other needs. For example, it may be preferable to invest in water distribution or in additional water resources. There may be more benefits in supplying a larger number of people with a lower quality of water than in supplying a smaller quantity of high-quality water to a limited population.

This book concentrates on water treatment, only a small part of the water-supply system. It is, however, most important for water-treatment engineers, scientists, and managers not to adopt a blinkered approach but to look outwards and to consider the entire system from source to consumer, paying due attention to wastes produced. Chapter 2 addresses water quality. Subsequent chapters are devoted to treatment of water and wastes from water treatment. In order to put water treatment in context, Chapter 14 considers some of the factors relating to water demands and demand management, an increasingly important area as many countries approach the point at which additional water resources become scarce and increasingly difficult and expensive to exploit.

2: Quality of water

ORIGIN AND TYPE OF IMPURITIES

Absolutely pure water is never found in nature and it is increasingly rare to encounter a source of water that requires no treatment before being used for potable-water supply. Water contains both biological and inorganic matter. It is normal to classify the impurities found in water in one of three progressively finer states—suspended, colloidal, and dissolved. The method of treatment required for the removal of impurities, or their reduction to acceptable limits, depends in part on the fineness of the material. While the matter found in raw water may render it unfit for human consumption, treatment can also adversely affect water quality by introducing pollutants, or by modifying chemicals that are harmless prior to water treatment.

Suspended matter

Running water obviously will carry floating debris, but it also has the capacity to pick up and transport solid particles of greater density than water; the higher the velocity the bigger the particle that can be transported. Rivers are normally at their most turbid during flood, because of the increased water velocity. Table 2.1 indicates the sizes of solids that are transported at different velocities. Large rivers in flood often run at velocities in excess of the maximum shown in Table 2.1 and are capable of carrying high quantities of suspended material.

Colloids

Colloids are fine particles that do not settle and which are electrically charged. The particles have a similar electrical charge, normally negative, which prevents them from coalescing together to form

Table 2.1. Transportation velocities of particles (adapted from Fox[1])

Material	Diameter of particle (mm)	Velocity of water (m/s)
Fine sand	0.4	0.15
Medium sand	1.1	0.23
Coarse sand	2.5	0.3
Gravel	2.5–25	0.76
Shingle	25–75	1.2

larger settleable particles. They are invisible to the naked eye, but can impart colour and turbidity to the water.

Dissolved solids

In its passage over or through the ground, water may dissolve a wide variety of chemicals. Common cations are aluminium, calcium, sodium, potassium, iron, and manganese. Common anions are bicarbonate, chloride, sulfate, and nitrate. A soft upland water will typically have a total dissolved solids (TDS) of between 70 and 150 mg/l and a hardness of 30–100 mg $CaCO_3$/l. Lowland rivers have a TDS typically of between 200 and 400 mg/l and a hardness of 100–250 mg $CaCO_3$/l. Groundwaters can have a TDS of up to 500 mg/l or even higher and a hardness of 400 mg $CaCO_3$/l or more. Gases also may be present in groundwaters, particularly carbon dioxide (CO_2), oxygen, nitrogen, and ammonia. Naturally occurring dissolved solids (DS) are rarely objectionable in low concentrations.

As well as DS derived from minerals, water may also dissolve man-made pollutants notably agricultural chemicals such as pesticides or herbicides. The levels of these normally vary over the year, dependent on times of application and rainfall/runoff. Although concentrations of these chemicals are typically very low, they can be very expensive to remove.

Organic pollution

Pollution from organic matter can be serious, particularly for groundwater sources which receive little treatment. Faecal pollution, whether from animals or humans is of particular concern, given the risks of disease transmission. The presence of ammonia, nitrates, and nitrites, which are products of decomposition of organic wastes, indicates the possibility of faecal contamination. The presence of bacterial indicators is taken to confirm this.

Water derived from peaty upland catchments can have high levels of organic colour, often associated with humic and fluvic acids. It is often difficult to treat such waters, and chlorination can lead to the production of trihalomethanes (THMs). Waters containing algae may also be prone to high levels of THMs after chlorination. Chapter 12 discusses the formation of THMs due to chlorination. Chloroform is the most common THM.

Algal toxins

Where water is extracted from lakes or reservoirs, there is a risk of the presence of algal toxins produced by blue–green algae (cyanobacteria). There are three classes of toxins potentially of drinking water significance:

- hepatotoxins—which affect the liver;
- neurotoxins—which affect the nervous system may lead to respiratory arrest; and
- lipopolysaccharides—some of which are irritants.

Dermatotoxins are also produced which may cause severe dermatitis to swimmers. The greatest concern is compounds of the microcystin family. These are produced by several species of blue–green algae and have been implicated in several incidents of liver problems, mainly in Australia.[2]

Microbiological parameters

The greatest short-term threat to human health from drinking water derives from pathogenic micro-organisms. A prime objective of water treatment is therefore to produce water free from such organisms. One approach would be to analyse for specific disease-causing organisms. However, there are difficulties and risks in cultivating such bacteria, and it is not practicable to analyse directly for viruses. The approach has therefore been to look for the presence of easily identified bacteria that are known to be present in human faeces, and to treat their presence as an indication of possible faecal contamination. If these indicator organisms are not present, then the water is assumed to be free of human pathogens and suitable for human consumption.

Traditionally there have been three or four main bacteriological parameters used to assess the quality of drinking water and to alert water suppliers and the quality regulators to the potential presence of organisms pathogenic to humans. The parameters which are normally analysed are:

- coliform bacteria;
- faecal coliforms;
- colony/plate counts; and
- faecal streptococcus.

The 1998 European Union (EU) Drinking Water Directive[3] included a standard for surface waters for the sulfate-reducing bacterium *Clostridium perfringens*, which is another indicator of faecal pollution. British Standards also include a standard for *Cryptosporidium parvum* (known colloquially as crypto). This was introduced, because over recent years there has been increasing concern over the presence of protozoan oocysts, notably *Cryptosporidium* and *Giardia lamblia*, both of which cause severe gastroenteritis.

The potential transmission of viruses is a concern, but it appears that viruses are effectively inactivated by disinfection; and there is little evidence for viral transmission in properly treated water.

The significance of the above parameters is discussed further below. However, it is important to note that coliform bacteria are widely found in nature and do not necessarily indicate faecal pollution. The presence of either faecal coliform or faecal streptococcus is taken as evidence of faecal contamination. In practice, all surface water surface sources have coliforms present and most have faecal coliforms. However, groundwaters are often of high bacteriological purity. Plate counts give an indication of the overall level of bacterial activity in a sample and do not directly relate to faecal contamination.

Tastes and odours

Under certain conditions algae may be present in surface waters, particularly in water from reservoirs or lakes. These can lead to objectionable tastes and odours in treated water; these are particularly associated with chlorination. The removal of algae is essential and often difficult. Certain other substances may also cause taste even when present in incredibly low concentrations of as little as 1 μg/l. After chlorination, phenols may give rise to taste from chlorophenols when present in concentrations down to as low as 0.01 μg/l.[4]

Hardness

The soluble salts of calcium and magnesium commonly found in water cause hardness. Hardness forms insoluble precipitates with soap and requires more soap to be used to obtain lather. It also causes boiler scale. In the past, it was common to soften hard waters. However, there is a view that soft waters are associated with heart disease and thus softening is less common now. However, in the UK there is sometimes a legal requirement to soften supplies to specific areas; such requirements are often associated with Acts of Parliament used in the past to promote water-supply schemes. Hardness and water softening are discussed in more detail in Chapter 11.

Iron and manganese

Iron and manganese impart colour, and can lead to staining of washing. Iron may be derived from raw water or from corrosion of iron water mains. Manganese is derived from raw water. These elements are common in some groundwaters, and may also occur in water taken from the lower levels of reservoirs. In the UK, both iron and manganese are removed by water treatment where necessary. Thus, the presence of iron in drinking water is therefore normally associated with corrosion of the distribution system. Where iron and manganese are found in drinking water, the cause is normally disturbances to the distribution system causing re-suspension of old deposits; typically such events are due to pipe bursts, work on the distribution system involving changes in flow patterns, or the use of fire hydrants by the fire brigade or others.

Sulfates, chlorides, bromides, and fluorides

The sulfates of magnesium and sodium if present in excess act as laxatives. Chlorides in concentrations above 600 mg/l tend to give the water a salty taste. Bromides have only recently become recognized as a potential raw-water-quality problem; this is because when water-containing bromide is ozonated, there is a risk of bromates forming, for which the allowable concentration is very low. Fluorides in low concentrations of approximately 1.0 mg/l provide protection from tooth decay and fluoride is dosed in some areas for this reason. However, levels above 1.5 mg/l are undesirable and may affect bones and cause mottling of teeth.

Corrosiveness and plumbosolvency

Waters with high CO_2 concentration, low pH value, and low alkalinity are generally corrosive. This is of importance as corrosive water can attack materials used in the distribution network and domestic plumbing systems, and can lead to elevated concentrations of iron, copper, and lead. A commonly used measure of whether water is corrosive is the Langelier saturation index (LSI). The LSI is the difference between the actual pH value of the water and the pH value at which the water would be saturated with calcium carbonate (pH_s). It therefore indicates whether particular water will dissolve or precipitate calcium carbonate. If the index is positive, the water will precipitate calcium carbonate, tending to protect pipework from corrosion. However, while the LSI is useful it is only a very crude indicator of potential corrosion problems. Corrosion is far more complex than this.

   Plumbosolvency refers to the tendency of a water to dissolve lead, which is present in many older domestic plumbing systems, in lead pipes, in lead joints in cast iron pipes, and in the solder used to join copper pipes. Excessive lead concentrations are associated with a range of health problems, particularly in infants and young children, including a reduction of IQ,[5] and as a result there is a requirement to greatly lower lead levels in water. Because of the tightening of lead standards in the 1998 EU Drinking Water Directive, pH value adjustment and the dosing of phosphate-based chemicals to reduce corrosion has become very common. Among natural waters, soft, peat-stained moorland waters are generally corrosive but many other waters also require treatment to reduce plumbosolvency in order to meet the lead standards.

RIVER AND SOURCE QUALITY

In the early years of public water supply, many towns and cities established upland reservoirs within protected catchments. This ensured that the quality of the water in the reservoirs was bacteriologically good and that water from the reservoirs could be supplied at low risk directly to consumers. As water-treatment technology developed, the availability of potable-quality water at a source was seen to be less important, and abstraction points moved steadily downstream, but with full treatment. The development of water-quality standards has led to standards for many organic chemicals found in low concentrations. This has led to increased treatment costs and a renewed appreciation of the importance of high-quality sources for drinking water. It has also led to the realization that it can be more economic to control the pollutants entering the aquatic environment than to treat water to remove them. Thus, the use of a herbicide that is less effective, but that rapidly and harmlessly degrades, may have a lower overall cost to society and may be preferable to a more effective, but persistent, alternative that requires expensive water treatment to remove it. This sort of calculation is increasingly being done in the regulation of agricultural and other man-made chemicals.

River quality can be assessed by chemical-based parameters, including biological oxygen demand, and by biological indicators that take into account the presence of indicator organisms and the diversity of life found in water. A 1975 EU Directive concerning the quality of surface water intended for the abstraction of drinking water in the Member States (75/440/EEC) classified surface waters into three classes. It specified that:

- Category A1 water requires simple physical treatment and disinfection;
- Category A2 water requires normal physical treatment, chemical treatment, and disinfection; and
- Category A3 water requires intensive physical and chemical treatment, extended treatment, and disinfection.

Classification is mainly on the basis of chemical and microbiological quality, without any biological indicators. It was incorporated within UK law for England and Wales by the Surface Waters (Abstraction for Drinking Water) (Classification) Regulations 1996, and is referred to in the 2000 Drinking Water Regulations as the basis for the treatment required for a particular surface water source.[6]

Water Framework Directive

In October 2000, the EU introduced a Framework Directive (2000/60/EC) to cover EU policy in relation to all waters, including groundwater, rivers, estuaries, and coastal waters. It sets environmental objectives and a target date of 2016 to achieve them. Member States have to prepare river basin management plans for all basins. The new Directive repeals seven existing Directives including that referred to above which classifies waters used for drinking water abstraction. It is not possible to foresee what the full implications are for water supply, but two areas stand out: there will be an improvement in the quality of some rivers, making them more suitable for use for public water supply; and the Directive could lead to some restrictions on abstractions where they adversely affect the quality or ecology of waters.

POTABLE-WATER STANDARDS

For much of the twentieth century the key requirement for water quality in the UK was that the water should be wholesome. In practice, this meant that the key standards were that the water be free of coliforms and *Escherichia coli*. However, development of prescriptive water-quality standards, notably by WHO, covering a range of physical- and chemical-quality parameters for drinking water, started in the 1950s and still continues. Additional parameters have been added as knowledge was gained on the health effects on humans of chemicals in water, and also as analytical techniques developed, allowing the identification of large numbers of organic chemicals in water.

WHO Standards

The first edition of WHO's *International Standards for Drinking Water* was published in 1958, and the second edition in 1963. The 1963 standards set coliform standards for microbiological purity, health-based maximum allowable concentrations for seven metals, and maximum acceptable and maximum allowable concentrations for a further 18 parameters. They also considered acceptable limits for fluoride and nitrate, without setting any standards, and suggested radiological limits.

In 1984 and 1985, WHO first published *Guidelines for Drinking Water Quality*. These stepped back from setting maximum allowable concentrations, rather they proposed guideline values to ensure aesthetically pleasing water posing no significant risk to health. The guidelines were published in three volumes, with the basis of the guideline values made clear in the supporting data. The introduction to the guidelines made it clear that the first priorities were to make water available and to ensure its microbiological quality. The 1984 Guidelines propose guideline values for 43 parameters.

In 1993, Vol. 1 of second edition of the *WHO Guidelines* was published; updated by an Addendum published in 1998.[5] This recommended new guideline values, again based on the need to provide aesthetically pleasing water providing no significant risk to health. The number of parameters for which guideline values were given increased to around 140, including approximately 90 organic chemicals of health significance.

WHO are preparing a new edition of their water-quality guidelines and publication is expected in 2003. The new edition is expected to be published on the Internet, and to be subject to regular rolling updates rather than periodic re-issues.

The 1963 WHO standards were of great importance in the development of water-quality standards in much of the world. They were arguably the first standards widely used by engineers and governments to set quality standards for new water-treatment plants and as the basis of national water-quality legislation.

European Standards

In the UK and the EU, the defining moment in the development of prescriptive water-quality standards was the adoption of the 1980 Drinking Water Directive (80/778/EEC). This set the minimum water-quality standards to be adopted by the member countries, a population of over 350 million. It also set sampling and analytical frequencies, to check whether the standards were being met. The Directive covered 62 parameters, divided into microbiological, toxic substances, substances undesirable in excessive amounts, physicochemical parameters, and organoleptic parameters. For the majority of the parameters, it set both guide levels and maximum allowable concentrations, but for toxic substances and microbiological parameters only maxima were set, and for some parameters only guide levels were given. The Directive allowed limited latitude to national governments in respect of relaxations of standards, although governments were able to impose more stringent standards. The Directive applied to public water supplies, and not to bottled water.

The 1980 Directive was responsible for an enormous improvement in both the quality of drinking water across Europe and in the data available on drinking water quality. It was also responsible for a vast investment to meet its standards. The 1980 Directive was the subject of some criticism; notably, that it over-regulated some harmless substances, such as silica; that it set some standards, notably pesticides, based on ideals rather than an assessment of health risks; and that it set statistically unsound absolute standards. More recently, it became increasingly apparent that it was too lax for some parameters, notably lead. By the early 1990s, pressure was mounting for the Directive to be revised and after much discussion a revised Directive was adopted.

The revised Directive on the quality of water intended for human consumption was adopted in 1998 (98/83/EC).[3] The new Directive adopts the principle of subsidiarity, focusing on essential health and quality parameters and allowing a degree of latitude on other parameters. The Directive divides the quality parameters into three parts: part A sets mandatory microbiological parameters, part B sets mandatory chemical parameters, and part C sets indicator parameters, chemical, physical, microbiological, and radioactivity. Part C contains many parameters that were previously mandatory. The part C parameters are set for monitoring purposes and a failure to meet these standards need not require remedial action unless there is a risk to human health. The Directive allows only a limited period for derogations (effectively relaxations of the mandatory standards).

The 1998 Directive represents a dramatic change in focus. The 1980 Directive led to a large and ongoing investment to meet the standards for iron, manganese, and aluminium. Under the new Directive, all this investment would not have been mandated. The 1998 Directive allows European governments to relax water-quality standards in some areas. However, it seems unlikely that this will be politically acceptable. In other areas, there is a significant tightening of standards, most notably for lead and bromate. The 1998 Directive covers bottled water as well as piped supplies. Table 2.2 compares the old and new European Drinking Water Directives for some key parameters and Appendix 2 provides more detail.

UK and Ireland Regulations

There was a significant delay in incorporating the requirements of the 1980 Directive into the law in many countries. In the UK,

England and Wales were covered by the 1989 Water Quality Regulations; Scotland was covered by the Water Supply (Water Quality) (Scotland) Regulations 1990; and Northern Ireland by the Water Quality Regulations (NI) 1994. The equivalent Republic of Ireland regulations are the 'European Communities (Quality of Water Intended for Human Consumption) Regulations 1988 (SI No. 81 of 1988)'. The UK regulations tightened the EU standards in some areas, specifically for lead and THMs. The Regulations referred to above will be superseded in 2004 by Regulations based on the 1998 EU Directive.

The requirements of the 1998 European Directive have now been incorporated into British and Irish law. In England and Wales, this has been done by 'The Water Supply (Water Quality) Regulations 2000'.[6] The UK Regulations maintain most of the standards set out in the earlier regulations, but add some additional parameters derived from the 1998 European Directive, and tighten some others. Notable changes are the tightening of the lead standard, to 10 µg/l in 2013, and the introduction of standards for bromate, nitrite, and some polymers used in water treatment. The Republic of Ireland introduced the European Communities (EC) (Drinking Water) Regulations 2000 (SI No. 439 of 2000) to implement the 1998 EU Directive.

As well as the standards set out in the Regulations, there are also paragraphs in the UK Regulations which require materials and chemicals used in drinking water supply to be approved. In England, this is done by a list of approved products and processes.[7] This sets standards for the concentrations of some chemicals arising from water treatment additional to those in the Regulations. An example is the limit on by-products of chlorine dioxide generation. In England and Wales, this list is derived from the requirements of Paragraph 25 of the 1989 Regulations and, from 2004, of Paragraph 31 of the 2000 Regulations.

USA Standards

The US Standards are the current National Primary and Secondary Drinking Water Regulations.[8] Primary standards are legally enforceable standards that regulate contaminants that can adversely affect public health. Secondary standards relate to aesthetic parameters and are non-enforceable Federally; however, individual states may adopt them as enforceable standards. The US Standards include the concept

Table 2.2. *Comparison of WHO, European, and USA drinking water-quality standards (all values in µg/l unless otherwise indicated)*

Selected parameter	WHO Guideline value (1993) (Note 1)	EU—1980 (80/778/EEC)	EU—1998 (98/83/EC) (Note 7)	US Federal Standards (state standards may be tighter)
Turbidity (NTU)*	5	4	Acceptable (C)	Note 6
Colour (Note 3)	15 (Note 3)	20 (Note 3)	Acceptable (C)	15 (Note 4)
Lead	10 (health)	50	10 (B)	15 (Note 5)
Iron	300	200	200 (C)	300 (Note 4)
Manganese	100 500 (health)	50	50 (C)	50 (Note 4)
Aluminium	200	200	200 (C)	50–200 (Note 4)
Cadmium	3 (health)	5	5 (B)	5
Arsenic	10 (health)	50	10 (B)	50
Nitrates (as NO_3)	50 mg/l (health)	50 mg/l	50 (B)	44
Sulfates	250 mg/l	250 mg/l	250 (C)	250 (Note 4)
Bromates	25 (health)	—	10 (B)	None
Cyanide	70 (health)	50	50 (B)	200
Lindane	2 (health)	0.1 (Note 2)	0.1 (B)	0.2
Aldrin	0.03 (health)	0.1 (Note 2)	0.03 (B)	—

Vinyl chloride	5 (health)	—	0.50 (B)	2
Bromoform	100 (health)	—	Total THM 100 (B)	—
Chloroform	200 (health)	—	Total THM 100 (B)	Total THM 100

*NTU: nephelometric turbidity units.

Notes:

1. For WHO Guideline values the table indicates if the parameter has been set for health reasons. All other parameters are for aesthetics.
2. All individual pesticides have a limit of 0.1 µg/l. The total concentration of all pesticides must not exceed 0.5 µg/l.
3. 1980 Directive units in mg/l Pt/Co scale. WHO units in true colour units (TCU).
4. Secondary standard—non-enforceable although individual states may have as an enforceable standard.
5. Action level—if exceeded by more than 10% of samples, treatment is required.
6. Surface water or groundwater under influence of surface water requires filtration such that no sample should exceed 5 NTU and 95% of samples must be below 1 NTU (or 0.5 NTU for conventional or direct filtration).
7. 'B' indicates mandatory chemical parameter, and 'C' indicates indicator parameter.

of a legally enforceable 'treatment technique'; this applies to several parameters that it is difficult to set absolute standards for, including lead, copper, acrylamide (polymer), *Giardia lamblia*, and turbidity.

Comparison of standards

Table 2.2 compares the WHO, US Federal, and EU Water Standards for a selection of commonly found contaminants. UK Standards are based on the 1980 and 1998 European Directives. Appendix 2 contains further details of these standards.

ANALYSIS OF PARAMETERS

As well as leading to great improvements in water quality, the new regulations have led to a massive increase in the numbers of samples analysed. This together with increasing cost pressures has led to a transformation in analytical techniques over recent years. Nowadays, virtually all analyses are carried out by automated equipment.

KEY WATER-QUALITY PARAMETERS

Turbidity

Turbidity is caused by the presence of fine suspended solids in the water. It is a measure of the amount of scattering that occurs when light passes through water. It is not directly related to suspended solids. The latter is a measure of the total weight of dry solids present, whereas turbidity is an optical effect, stated in turbidity units, which also reflects the fineness, colour, and shape of the dispersed particles. For this reason, there is no constant linear relationship between the two.

There are several methods of measuring turbidity, and in the past the results were frequently expressed in different units. This led to considerable confusion, but nowadays NTU are almost always used. Turbidity values in NTU are for all practical purposes interchangeable with those expressed as Jackson turbidity units (JTU), formazin turbidity units (FTU), or the APHA turbidity unit.

Methods of measurement

There are a number of types of turbidimeter, most of which are now only of historic interest. In the past, measurement of turbidity meant using a Jackson turbidimeter. In this method, a standard candle is viewed through a column of water under test, the length of which is increased until the flame disappears. The length of the column defines the degree of turbidity, which is stated in JTU. Later, these were calibrated by using suspensions of formazin polymer with the results stated in formazin units. These units were widely used until superseded by instruments reading directly in NTU. NTU and JTU can be taken as numerically equal.

Turbidity is measured in NTU by directly measuring the scattering of light in a laboratory instrument or an on-line continuous monitoring instrument. These involve shining a light of specified wavelength through a sample of water in a cell through which water is continuously passed, and they immediately record turbidity. It is now normal practice to have several on-line turbidimeters at water-treatment works to continuously monitor raw, filtered, and treated water. A practical problem with many on-line turbidimeters is fouling of the cell through which the water passes; as the cell becomes dirty, the instrument becomes inaccurate. This can be avoided by shining the light through a falling column of water, avoiding fouling, but requiring precise control of the column of water.

Colour

Colour is a measure of the light absorbed by the water. True colour excludes the effects of any scattering of light due to turbidity. Colour in drinking water is normally measured by a comparison method. This defines a unit of colour, equal to a °H, as that produced by 1 mg of platinum/l in the form of the chromoplatinate ion. Colour is measured by comparing the sample against a set of standard colour solutions. There are also automated instruments which measure colour directly. Normally, colour is very pH dependent and it is good practice to also report pH value.

There is also a spectrophotometric method that measures colour more precisely in terms of light wavelengths; but this is not used in water treatment, except for research.

Pesticides and other organic chemicals

The standards for many organic chemicals have been derived in parallel with the development of powerful analytical techniques.

As techniques have developed, more and more chemicals have been found to be present in water, and standards have been developed to cover these chemicals. The 1980 EC Drinking Water Directive specified a blanket limit of 0.1 μg/l for all pesticides; this was intended to be effectively a zero limit. In 1980, the standard was below the level of detection for many pesticides, and led to the development of analytical methods to measure pesticides at such low concentrations. These techniques were also appropriate to the analysis of many other complex organic chemicals.

The standard methods for analysing many of the organic chemicals present in water at μg/l concentrations involve extraction and concentration of the chemical from water followed by gas chromatography or mass spectrometry. The equipment is complex and these analyses are carried out using automated equipment.

MICROBIOLOGICAL PARAMETERS

Coliform bacteria

These are rod-shaped bacteria which are widely found in the natural environment. Water-quality testing is particularly interested in those bacteria that may be derived either from the human gut or from the gut of other warm-blooded animals, which thus prosper under the conditions found in the human digestive system. This is an acid environment at a temperature of approximately 37°C.

Coliform bacteria are defined based in part on the method historically used to detect them. They are rod-shaped bacteria that ferment lactose, forming gas and acid, in the presence of bile salts at a temperature of 35–37°C within 48 h. The old method of analysis used multiple dilutions of the water being tested, which were incubated in tubes. The number of positive tubes at the different dilutions gives the most probable number (MPN) of bacteria present. The result was expressed as a number of coliforms per 100 ml. The test was labour-intensive and has largely been superseded by methods using membrane techniques that filter out bacteria and then grow colonies of coliform bacteria, allowing a direct estimate of the number of coliforms present in a sample. However, these membrane techniques are in turn being replaced by enzyme substrate methods that permit simple analyses of both coliform and E. coli organisms; these are discussed further below.

Escherichia coli (E. coli)

If coliform bacteria are present, it is necessary to distinguish those that may derive from other sources from those that could only derive from the gut of warm-blooded animals. The *E. coli* test confirms that the coliforms are of faecal origin. *E. coli* or faecal coliforms are bacteria that can produce gas from lactose at a temperature of 44.5°C. Their presence is taken as confirmation of faecal contamination. Again both multiple tube methods and membrane filtration methods can be used, and both methods are likely to be replaced by enzyme substrate methods for routine analyses.

As coliform bacteria can arise from sources other than human faecal contamination, the presence of such bacteria is permissible in a small proportion of samples, normally no more than 5% of samples. The presence of *E. coli* is unacceptable and none should be detected in public potable-water supplies. Where coliform bacteria are found, it is usual to then check whether they are *E. coli* by additional analysis.

Enzyme substrate analyses (Colilert®)

The method involves the use of commercially available substrate formulations that are used as a substrate by coliforms and *E. coli* bacteria. If coliforms are present, a yellow colour is produced, and if *E. coli* are present, a fluorescent chemical is also produced. The formulation is added to dilutions of the water being tested, which are then incubated and examined under normal and fluorescent light to confirm the presence of coliforms and *E. coli* bacteria. The normal procedure is to put the water and reagent into a plastic disposable tray containing a large number of cells. The tray is sealed and incubated for around 24 h. By counting the number of positive cells, either yellow or yellow and fluorescent, the number of coliforms and *E. coli* can be estimated using MPN techniques.

Such methods allow rapid and easy enumeration of bacteria and they are beginning to be used for day-to-day monitoring of drinking water quality by water companies. The great advantage of this method is that, if coliforms are present, it avoids the need for further confirmatory analyses to test for *E. coli*.[9]

Plate counts

Plate counts examine the total number of viable bacteria by mixing a sample of water with a medium and then allowing the mixture to incubate at a temperature of either 22°C or 37°C. The plate will develop colonies, indicating the number of viable bacteria and these can be counted. The test is useful for detecting changes in microbiological quality. Values found in tap water will vary from place to place but are normally consistent and low, typically 10–100 colonies/ml. Changes can indicate pollution of the distribution system, lack of chlorine, or a water-treatment failure. No standards are normally set for plate counts, but in the UK they are a required test on statutory water-distribution samples.

Faecal streptococci

These are species of bacteria that have been identified in the guts of warm-blooded animals. A particular subgroup is the enterococcus group that is referred to in the EC Water Directive. The prefix 'faecal' indicates that they can thrive in conditions present in the human gut. They are generally accepted to be more persistent than *E. coli* and coliform bacteria, and may indicate remote pollution. Faecal streptococci can grow at 45°C in concentrations of bile and sodium azide which inhibit coliform organisms. Analytical methods are based on this. Membrane filtration methods are mainly used nowadays, although enzyme substrate methods have been developed and are likely to be used in the future.

Cryptosporidium parvum

In 1999, the UK introduced a quality standard for a protozoa, *Cryptosporidium parvum*.[10] This is a very persistent protozoa that gives rise to severe diarrhoea which cannot be medically treated. It is extremely difficult to detect and determine the concentration of the organism in water. The standard is one organism in 10 l. Water-treatment works considered at risk have to effectively monitor continuously, taking one sample per day continuously over 22 h. A sample of water is filtered in a cartridge filter to remove the *Cryptosporidium* oocysts. The volume of water passed through the filter over the 22 h period is measured and has to be at least 1500 l. Laboratory analysis involves extracting the oocysts from the filter in a small volume of water. The oocysts are recovered from this water and concentrated by a complex process involving further filtration, centrifuging, and use of specific magnetic antigens, eventually concentrating the oocysts onto a slide. After staining, the slide is then examined under the microscope and oocysts are counted. The process is complex and labour-intensive. Present analytical methods are believed to count up to 40% of oocysts present. In the UK, it costs around £50 per analysis, an annual cost of around £20 000 per site at which monitoring is required.

The current test (and the standard) does not differentiate between viable and non-viable oocysts, and can also extract other non-pathogenic organisms of similar size and appearance. Given the large amounts of money being spent on crypto-monitoring, it is likely that better analytical methods will be developed over the next few years.

CONCLUSIONS

While the development of water-quality standards over the past few decades has led to greatly improved water quality in the richer countries, it can also be argued to have had some negative impact. In the past, there were no hard and fast rules as to the acceptable quality for potable supplies, but clearly the water had to be wholesome. The position now is very different over much of the world; usually there are now extensive standards. In the richer countries, this has led to a vast investment in water treatment to meet the often very strict water-quality criteria. In poorer countries, the development of water-quality standards has arguably diverted funds into a limited number of modern-treatment plants, while water in large areas of many poorer countries receives only very limited treatment and water fails to meet even the most essential microbiological parameters. The introduction to the 1993 WHO *Guidelines for Drinking Water Quality* stresses the importance of prioritizing water-quality standards and the absolute priority that should be given to parameters affecting public health. However, aesthetic parameters are also important in that if water is unpalatable, consumers may use alternative unsafe but more palatable supplies.

A very significant proportion of the recent investment in water treatment in the UK has been to meet the standards for organic chemicals introduced into the aquatic environment by human activity, most notably pesticides. A figure of $2.25 billion was cited[11] in

1990 as the cost per life saved in meeting the USA drinking water standard for the pesticides simazine/alachlor. While this figure may well be inaccurate, it does highlight two important points: it is very expensive to treat for some chemicals once they are in the environment; and over much of the world meeting such standards would represent a misallocation of resources.

3: Overview of water treatment

INTRODUCTION

Arguably it is possible to produce potable water from virtually any source of water. In practice, of course, there are a number of restrictions on the quality of water that can be treated. The usual reason for not using a particular source is cost but there are other reasons, notably ecological/environmental constraints and aesthetics—direct potable re-use of wastewater is usually unacceptable at present, although it has been practised at Windhoek in Namibia since the early 1980s. Although ecological and environmental constraints can also be valued, aesthetic objections tend to have a value above mere money.

There is a whole range of treatment processes that can be used to treat a particular water. The actual processes selected will depend on the nature of the raw water, the space available to construct the plant, a consideration of operating costs and capital costs, and, often, personal or company preferences. There may well not be a single correct process train to treat particular water, but a number of possible options. Often engineers are further restrained by the need to upgrade an existing plant, resulting in further limitations on the processes that can be used.

This chapter considers water treatment under three main headings: traditional; modern; and the future. There is a brief consideration of developing world treatment, highlighting some of the different considerations. The classification of processes under these headings is obviously subjective; reflecting the authors' age and experience, but it does offer a convenient way to sub-divide the various processes. In practice water suppliers increasingly provide what is perceived to be the most economic treatment process within the existing physical constraints, taking account of risks and reducing them to an acceptable level. Mixes of traditional and advanced treatment will be provided to treat particular problems. One important issue is how risks are handled; it may be acceptable to a particular company or country to design a plant with a higher risk of failing particular quality standards than to incur the additional costs of reducing the risk. Considerations of this sort may be thought of as largely academic in the UK, where quality of the treated water is paramount, but in poorer parts of the world they are most important.

TRADITIONAL WATER TREATMENT

What processes can be classified as traditional water treatment? For surface waters conventional water treatment consists of some or all of the following main processes:

- catchment control;
- raw-water storage;
- removal of coarse solids by screening;
- sedimentation;
- aeration (also common for groundwaters with high levels of iron);
- chemical dosing;
- coagulation and flocculation;
- clarification;
- slow sand filtration;
- rapid gravity filtration;
- pressure filtration (also common for groundwaters);
- chlorination (also common for groundwaters);
- chlorine contact tank (also common for groundwaters);
- settlement and recycling of filter washwater;
- disposal of sludge to lagoons; and
- dewatering of sludge.

For groundwaters, the main problems tended to be elevated levels of iron and manganese and most groundwater treatment plants treated for these parameters. Less commonly groundwater treatment plants softened waters.

In the past, the core of most water-treatment plants was settlement/clarification basins and filters. They arguably remain the most important processes, but with the development of membrane-based processes their use is no longer essential.

Most water-treatment works constructed in the UK before the 1970s were unlikely to have processes other than those listed above. The particular process used varied from works to works and there were often important differences between works using similar processes. For example, rapid gravity filtration could be conventional sand filtration, dual media filtration, or even upward-flow filters. The processes are predominantly, although by no means solely, physical. For many relatively unpolluted sources such treatment produced water of acceptable quality, reducing turbidity, colour, suspended solids, and iron and manganese to acceptable levels and producing bacteriologically acceptable water.

Slow sand filtration became important partly as a result of the Metropolis Water Act of 1852. This prohibited the extraction of water from the tidal Thames and required that all surface-derived water supplied to London should be treated by sand filtration. This requirement gave an enormous boost to the construction of water-treatment plants to serve London. Despite this, it is surprisingly recently that there has been a move towards universal treatment of water, other than by disinfection. At the time of writing there are still cities in the British Isles (for example, Belfast and Manchester) that receive some surface water from protected catchments that has essentially had no treatment other than disinfection, and in the USA, most of New York's water is untreated save for disinfection, fluoridation, and corrosion control.[1]

The supply of untreated water was predicated on the idea that collecting water from controlled catchments would ensure a safe water of satisfactory quality. Thus catchment control involved limiting human access to, and prohibiting most agricultural activity in, the catchment. The view that catchment control alone provides water that is safe after disinfection is no longer acceptable; there have been too many outbreaks of waterborne diseases associated with protozoan parasites, specifically *Giardia lamblia* and *Cryptosporidium parvum*. Furthermore, the introduction of absolute quality standards for colour, turbidity, iron and manganese means that untreated water may be unacceptable on account of failure to meet these predominantly aesthetic parameters. In addition there is increasing pressure for recreational access to protected catchments. The supply of surface water untreated save for disinfection will become very uncommon by 2020.

MODERN WATER TREATMENT

In parallel with the development of water-quality standards, water-treatment processes underwent a period of development. In the UK and Europe the adoption into national laws of the water-quality standards in the 1980 EU Drinking Water Directive led to the introduction of new treatment processes and the further development of traditional treatment processes to meet these standards.

Modern treatment processes include:

- improved coagulation control;
- dissolved air flotation (DAF);
- advanced clarifiers (lamella separators and advanced 'sludge-blanket' systems);
- ozonation;
- granular activates carbon (GAC) adsorption;
- membrane-based processes;
- air stripping of volatile organic chemicals; and
- advanced disinfection (ultraviolet, ozonation and chlorine dioxide).

Most of the above processes are used to deal with organic chemicals present in water, either naturally occurring compounds that give rise to colour, taste, or chlorination by-products, or pollutants arising from human activities, notably pesticides and herbicides. In addition, the introduction of computer-based systems has led to great improvements in the monitoring and control of the traditional processes, leading to higher and more consistent quality of water produced from traditional plants.

OVERVIEW OF PROCESSES AND PROCESS SELECTION

Table 3.1 lists the main current water-treatment processes and some specific water-quality problems for which the processes may be used. The table indicates the processes that may be applicable for specific water-quality problems. Typically for any water-quality problem there are options that can be used, depending on particular circumstances.

This table is not intended as a guide to process selection as the real world is more complex. It is always essential to approach water treatment with an open mind, looking at the particular problems of the water to be treated; the amount of money available; plants treating similar water; and, where a plant is to be upgraded, the performance of the existing plant. It is also often essential to consider carefully the best source of the water to be treated if the most economic solution is to be proposed; it may be that by carrying out

Table 3.1. Main water-treatment processes and water-quality problems addressed

Process	* Quality problem													
	1	2	3	4	5	6	7	8	9	10	11	12	13	14
Coarse screening	x													
Fine screening	x													
Microfiltration	x											x		
Raw-water storage		x								x				
Preliminary settlement		x												
Aeration					x	x								
Air stripping									x					
Coagulation and flocculation			x	x		x				x		x		
Gravity clarification			x	x		x				x		x		
DAF			x	x		x				x		x		
Slow sand filtration			x	x		x				x		x		
Rapid gravity filtration			x	x		x				x		x		
Microfiltration			x	x						x		x		
Ultrafiltration				x						x				
Reverse osmosis							x	x	x		x			
Activated carbon adsorption				x	x			x						
Pre-ozonation				x	x			x	x			x		
Post-ozonation								x	x					x
Ion exchange							x							
Chemical oxidation						x								
pH control						x							x	x
Phosphate dosing													x	
Chlorination						x								x

*1. Debris; 2. high-sediment load; 3. turbidity; 4. colour; 5. taste and odour; 6. iron and manganese; 7. nitrate; 8. pesticides; 9. volatile organic chemicals; 10. cryptosporidium; 11. salinity; 12. algae; 13. plumbosolvency; 14. microbiological quality.

works at the source or by modifying an intake the quality of the raw water can be improved to reduce the treatment needed.

TYPICAL PROCESS STREAMS

Given that there are many ways to treat a particular water, what processes are applied in practice? Figures 3.1–3.4 are typical process streams for a selection of waters: again these should not be considered as prescriptive.

Figure 3.1 represents the treatment water from the lower reaches of a large English river such as the Thames or the Severn. Such water requires extensive treatment to remove in particular the pollutants derived from upstream human activities and the treated wastewater that has been discharged upstream. Despite the number of processes shown, the water is normally relatively easy to treat.

Figure 3.2 is an alternative approach to Fig. 3.1, using bankside storage of raw water and slow sand filters.

Figure 3.3 represents a possible treatment process for water extracted from the upper reaches of a river draining upland moors. Such waters are likely to be low in pollutants derived from human activities, but could well be soft with high natural colour and be of concern with respect to *cryptosporidium* oocysts. These waters are difficult to coagulate, forming a light floc that does not settle well. Chlorination may lead to THM formation and is delayed as long as possible. If the source was assessed at high risk of animal-derived *cryptosporidium* oocysts being present in high numbers, the poor-coagulation characteristics of such waters may lead to the addition of an extra treatment stage specifically for *cryptosporidium* removal.

Figure 3.4 represents a possible treatment process for water extracted from boreholes or reservoir waters containing iron and manganese. The key problem is to remove both the iron and manganese, which require different chemical environments for most effective removal. The process shown includes two-stage filtration, but this is not always necessary, as discussed in Chapter 11.

RAW- AND TREATED-WATER QUALITY

Conventional water treatment is appropriate for most surface or groundwaters that are free of serious organic contamination derived from human activities, as well as for many waters that are not.

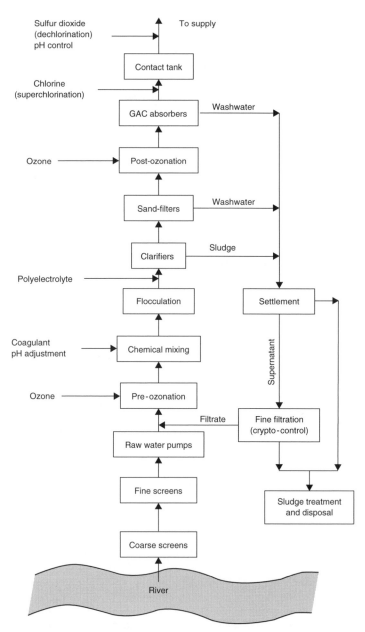

Fig. 3.1. Treatment schematic for lowland river water

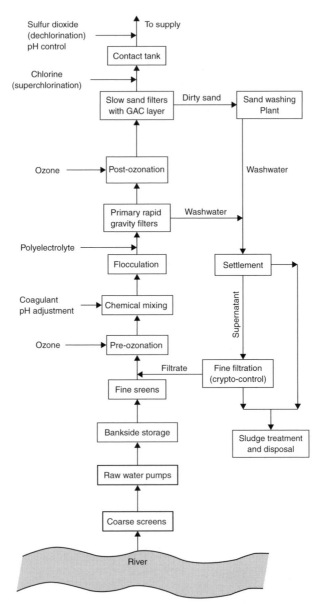

Fig. 3.2. Treatment schematic for lowland river water using slow sand filters

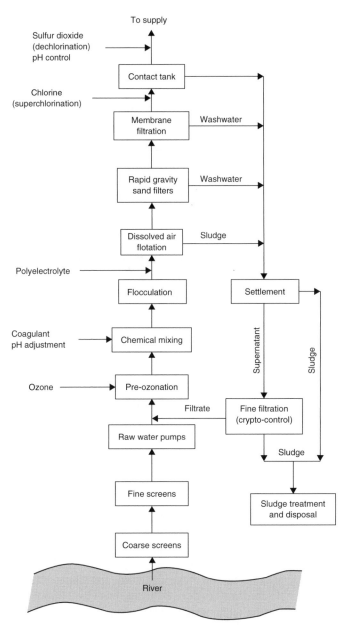

Fig. 3.3. Treatment schematic for upland river water

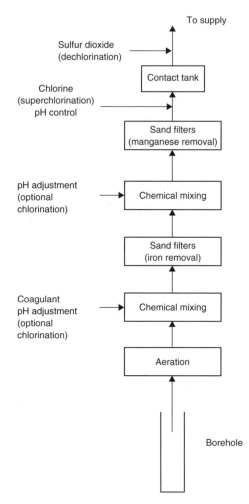

Fig. 3.4. Treatment schematic for borehole water

Conventional treatment can have difficulty treating some waters, notably soft-coloured waters derived from moorland and bogs. The difficulty of treating such waters has become more important as water-quality standards have developed. A conventional treatment plant might well have produced water that had long been considered potable, but might now no longer be acceptable because of, for example, a failure to comply with the colour or THM standards.

There have been many attempts to codify conventional water treatment, particularly for surface waters. These take into account key raw-water-quality parameters and define the treatment process required. Such broad approaches are of limited use. They may define a suitable starting point but often in practice the water engineer is either working within constraints that may severely limit the applicability of the solutions produced by such an approach, or the raw-water quality may be more complex than allowed for by a simple approach.

DEVELOPING WORLD TREATMENT

In the developing world there are important differences in priorities. Without doubt the overriding requirement is to provide a microbiologically safe water supply. Waterborne diseases are one of the greatest health hazards across much of the world and the high incidence of such diseases is one of the key factors causing high infant mortality rates in tropical countries with poor water supplies. This is not to say that all physical or chemical quality criteria should be ignored, but they should be considered with due regard for the overriding need for reliable supplies of microbiologically safe water at an affordable cost. There is no point in supplying safe water if it is aesthetically so poor, or so expensive, that other less safe supplies are used in preference.

The priorities in the developing world can be argued at length. A high priority should be to provide water from a source that requires little or no treatment, and to ensure the source is protected, rather than to use a lower-quality source and provide water treatment.[2] This ensures that when treatment fails, as it inevitably will, there is a far lower risk of waterborne disease transmission. A reasonable list of priorities is:

- wherever possible to utilize a safe source of water—often groundwater from a deep well—and ensure it remains protected;
- provision of effective disinfection, ensuring that the community is protected from acute waterborne diseases;
- ensuring a continuous supply of water;

- provision of an aesthetically acceptable water; and
- provision of high-quality water which ensures that the community is at low risk over a long period.

Where treatment is required, the first short-term aims are normally basic treatment to provide an aesthetically acceptable water and effective disinfection. This should ensure that the water supplied is safer than any alternative sources. Longer-term treatment objectives should be to comply with the health related standards set out in the WHO Guidelines for drinking water quality.

A concept encouraged by WHO[3] is that of multiple barriers, with respect to removal of pathogens. A single treatment stage will be prone to significant failures to supply water of an acceptable quality. By using several treatment stages a safer supply will be ensured. Important 'barriers' are storage, sedimentation/clarification, filtration, and reliable disinfection. It is vital that the failure of any single stage or barrier does not inevitably lead to unsafe water being supplied.

FUTURE TREATMENT PROCESSES

Looking into the future is more liable to result in future smiles than in accurate predictions. The current major water quality concerns arguably are lead, *Cryptosporidium*, and disinfection by-products.

Lead is at the time of writing a major operational and treatment concern for many European water suppliers. The lead standard is to reduce from the current level of 50 to 25 µg/l at the end of 2003, and then to 10 µg/l at the end of 2013. This is causing water companies to focus very carefully on the changes that often occur as water passes through a distribution system, and on precise control of chemical dosing and pH value control. A key problem with lead is often that the water put into a distribution system changes as it passes to customers. This may arise from biological action within the distribution system or it may be that the water entering the system is chemically unstable, or some combination of the two. Thus the pH value and alkalinity of the water may change. Also if phosphate is dosed to inhibit lead uptake, this may be lost as the water passes through distribution.

Lead control cannot be considered independently of other water-quality and -treatment issues. In particular there can be conflict between the pH value required for disinfection (less than approximately 8.0) and that required to minimize plumbosolvency which is often 8.0 or higher, and this may place constraints on where chemical dosing for pH value control can be applied. In the longer term the solution may be to remove lead from plumbing, but this is a major exercise that will take many decades.

It appears certain that there will be increasing use of membrane treatment in the future. The key drivers behind this in the UK are the standards for *Cryptosporidium* and THMs. Under UK Regulations the use of an approved 1 μm membrane filtration process removes the requirement to sample and analyse for *Cryptosporidium*. Such sampling and analyses significantly increases the cost of operating plants, particularly small plants that would otherwise require very little operating effort. The use of membranes to treat soft peaty upland waters is also becoming common. Such waters are coloured, difficult to treat, and often have unacceptable levels of THMs after chlorination.

Concern over disinfection by-products is likely to greatly influence water treatment. It has long been recognized that chlorination gives rise to the formation of THMs. More recently the formation of bromate as a by-product of ozonation has been identified as a major concern. There have been some concerns over chlorite levels where chlorine dioxide is used as a disinfectant in place of chlorine. In the future it is quite possible that other by-products or risks will be identified that will further restrict disinfection. It is possible, even though it seems almost inconceivable today, that the use of chlorine itself could be restricted.

Water treatment will continue to evolve, with more use of the modern treatment processes. However, it may well be that the most important changes over the period 2000–2020 will be the development of improved monitoring and control systems, covering the distribution system as well as the treatment plant. As the quality of water produced by treatment plants improves it will become unacceptable for it to significantly deteriorate in distribution systems. Thus black boxes filled with sensing and control equipment will become more and more important and will spread from treatment plants across the entire water network, and also to water sources, to allow treatment to be better optimized to variations in raw-water quality.

4: Preliminary treatment

Prior to the water passing to the main treatment processes there is normally some form of preliminary or pre-treatment. Processes classed as pre-treatment include raw-water storage, screening, aeration, straining, preliminary settling, and pre-ozonation. Any of these processes might be found at a particular plant but it is improbable that all would be needed. Each performs a particular function and unless the problem they are designed to eliminate is present in the raw water they can be omitted.

It used to be common to chlorinate water prior to treatment, to reduce problems of attached growth within the main treatment process but this is now uncommon due to the standards for THMs. Now it is increasingly common to pre-ozonate water, which disinfects, oxidizes some of the complex organic chemicals, and, for some waters, also improves the performance of the subsequent clarification process. Pre-ozonation is covered in Chapter 11.

COARSE INTAKE SCREENS

Coarse screens are provided at river intakes to prevent floating material of fairly large size entering the works. The steel bars forming the screen are normally quite substantial (typically about 25 mm dia. or equivalent) and are spaced about 100 mm apart. They are normally slightly angled from the vertical to facilitate raking. Sometimes the bars are mounted in frames which are duplicated so that one frame can be lifted for cleaning or repair without unscreened water entering into the plant. The velocity of water through the screen openings should not exceed 0.5 m/s. It is almost universal to provide coarse screening on intakes from surface water intakes. Most rivers at some time have floating or submerged debris that would damage pumps or block inlets. It is therefore prudent to always provide coarse screens. In some parts of the world there can be large and unpleasant items in the water, but in Britain it is most unusual to have to remove dead donkeys.

RAW-WATER STORAGE

The provision of raw-water storage on river intakes is good practice. There are several reasons why this is so. On intakes on the lower reaches of rivers there is always a risk that there will be significant upstream pollution, for example, a vehicle accident leading to diesel oil in the river or the discharge of a toxic chemical. To provide some flexibility in dealing with such incidents it is good practice to provide storage prior to a works. It is then possible to stop withdrawing water while the pollution passes down the river. In recognition of this risk the UK Government recommended in the early 1970s that a minimum of 7 days storage should be provided for treatment works the raw-water intakes of which were downstream of effluent discharges. The 7-day requirement for all works has not been achieved, and now never will be. However, 7 days is a reasonable period to allow for the worst pollution to pass and to provide time to deal with the source of the problem.

A significant improvement in quality of poor-quality water can be achieved by storage alone, even if such storage is not required for any other purpose. This improvement results from natural settling of the suspended solids and a marked reduction in pathogenic organisms. The improvement in water quality depends on residence time, whether the reservoir is fully mixed, and the time of year. Reservoirs with several weeks storage will normally give a reduction of over 90% in coliform and *Escherichia coli* numbers, rising to around 99% in the summer.[1] At the same time there will be reductions in colour, turbidity, ammonia, and many organic pollutants including pesticides and herbicides.[2] There will also be significant reductions in numbers of *Cryptosporidium* and *Giardia* oocysts. Table 4.1 presents some data illustrating the improvements that can be achieved. The storage in Grafham Water was of the order of 3 years.

Longer storage periods may be provided to balance the amount of water that can be abstracted from a river against demands, or to further improve the quality and consistency of the water supplied to a works for treatment. It is normal for abstraction licences to specify both the maximum quantity of water that may be abstracted and the

Table 4.1. Quality of water before and after storage

Parameter	River Thames at Oxford[3]		River Great Ouse at Grafham[4]	
	Raw water	Stored water	Raw water	Stored water
Colour (Hazen)	19	9	30	5
Turbidity units				
(type not specified)	14	3.2	10	1.5
Ammoniacal				
nitrogen (mg/l)	—	—	0.3	0.06
Oxygen adsorbed (mg/l)	1.8	1.3	3.5	2.0
BOD (mg/l)	—	—	4.5	2.5
Total hardness (mg/l)	300	259	430	280
Presumptive coliforms				
MPN/100 ml	60 000	200	6500	20
E. coli				
MPN/100 ml	20 000	100	1700	10
Colony counts per 1 ml				
3 days @ 20°C	—	—	50 000	580
2 days @ 37°C	—	—	15 000	140

minimum flow in the river downstream of the intake, with the latter taking precedence. Thus, for rivers with seasonal variations in flow it is often essential to provide large raw-water reservoirs to allow sufficient water to be abstracted at time of high flow to cover for the periods when abstraction has to be reduced at times of low flow. Storage periods shorter than 7 days will provide some protection against pollution but provide only limited other benefits.

Raw-water storage can also lead to problems, particularly where the water being stored is high in nutrients; this can lead to high levels of algae in the spring and summer. If the subsequent treatment process can handle this then there will not be a problem. If algae do cause problems then it will be necessary to take measure to limit their growth. Another problem can be silt deposition, which can be a particular problem on smaller reservoirs. While raw-water storage is undoubtedly beneficial, due allowance must also be made for potential problems in both the subsequent treatment processes and in the design of the storage itself, with provision for periodic access to remove silt and for managing potential water-quality issues. Storage equivalent to 7–15 days of the average water demand is sufficient to reduce pathogenic bacteria while being short enough to minimize problems associated with growth within the reservoir.

Large dams and deep storage reservoirs have their own water-quality issues. Problems arise from thermal stratification and there are often significant water quality changes over time in dams often associated with a steady build-up of nutrients and iron and manganese within the dam sediment over time. There are two main approaches to deal with these problems: draw-offs can be provided at different depths, allowing some control over the quality of the water drawn off; or a destratification system, normally using aeration, can be installed.

In Britain nowadays it is not easy to provide new raw-water storage. Where there are working or disused gravel pits in a river valley it may be possible to combine gravel extraction or rehabilitation of disused gravel pits with new bankside storage. In other areas this is not possible; land is expensive, there may be environmental objections to disturbing old gravel pits, and companies may be forced to protect against pollution risks by installing additional treatment rather than by providing storage.

ALGAE, ALGAL CONTROL AND RESERVOIRS

Algae are micro-organisms which contain chlorophyll a and require water, CO_2, low levels of inorganic substances and, most importantly, light to grow and multiply. They are essentially very small aquatic plants. They are slow growing compared to most bacteria, partly because they grow by photosynthesis. Their growth rates depend to a large degree on the availability of nutrients, nitrogen and phosphorus, in water, and on the energy available from sunshine. Thus, they tend to grow best in waters that have higher levels of nutrients. Because they grow relatively slowly they are more often associated with lakes and reservoirs rather than rivers, although long slow-flowing rivers may also have high algal concentrations. There are a multiplicity of types, many of which, if present to excess, can cause treatment problems at water-treatment plants. Algae can be either attached to the sides and bottoms of reservoirs or rivers, or can be free-floating. Blue–green algae, or cyanobacteria, are a form of algae with some similarities to bacteria. Algae can normally be seen in affected water, but some (notably *Stephanodiscus*) can remain invisible except under a powerful microscope. Outbreaks tend to be severe and sporadic. Algal blooms are a common phenomenon. These arise when the optimum conditions exist for a particular species of algae to grow. Algal growth is exponential and the species grows until there are large numbers present and growth is limited by lack of nutrients or by some other factor. The algae then die. Each year there tends to be a series of different species of algae which grow in large numbers in succession. A particular species 'blooms' and then dies; another species will then bloom and die, and so on. Algal blooms normally start in the spring, as increasing light levels lead to the beginning of optimum conditions for growth, and continue through the summer into the autumn. In an uncontrolled reservoir it is only in the winter that algal numbers are low for any extended period.

Algal blooms have a long history. It is possible that the plague of blood referred to in the Bible's Book of Exodus refers to an algal bloom in the River Nile. Nowadays they are extremely common across much of the world, brought on by increased concentrations of nitrogen and phosphorus in lakes and reservoirs. Algal growth and eutrophication of inland waters is recognized as a major problem and is now being addressed in EU countries. Nutrient removal is becoming increasingly common in sewage-treatment works that discharge to water bodies prone to algal blooms. However, a major source of nutrients is agriculture, and this may minimize any improvements due to nutrient removal from wastewaters.

In the past algal problems have sometimes come as a surprise to engineers; when the building of an impounding reservoir on a previously clear stream has lead to severe algal problems. Even when algae are not a problem with a new reservoir, as the reservoir ages there can often be a build-up of nutrients within the reservoir and over a period of time algal blooms may become a problem. Fairly alkaline waters containing appreciable concentrations of nitrates and phosphates are particularly suspect. Clarity is also a factor, as algae need sunlight for photosynthesis. Thus in rivers with a high silt load there are rarely problems with excessive levels of algae. However, in lakes and reservoirs in much of the UK and indeed much of Europe and North America, algae are a problem. It is generally only the remoter uphill lakes and reservoirs, which do not receive significant nutrient inputs from agriculture or sewage, that are free from large numbers of algae. Where a new dam is proposed the probability of algal problems can be assessed from the level of nutrients in the water, but it is foolhardy to assume that there will not be problems in new reservoirs.

Algal control is not easy. In the past it was common to control algal blooms by dosing copper sulfate before a bloom occurred. Copper is an algicide and the technique was effective in reducing algal blooms if properly monitored and applied. Copper was dosed at around 0.3 mg/l of copper sulfate. Except in very soft waters the copper was quickly removed by precipitation as copper carbonate. However, despite this, nowadays the dosing of relatively large quantities of copper into the aquatic environment is unacceptable because of its effect on other organisms. Copper sulfate is toxic to fish at a concentration similar to that needed to kill algae and thus fish kills were common. Algal growth is often limited by the availability of phosphorus and thus in smaller reservoirs, or reservoirs storing upland water low in phosphate, it may be practicable to limit algal growth by dosing a ferric salt to precipitate phosphate. However, this is relatively expensive and produces sludge that has to be removed periodically.

A problem with reservoirs more than around 10 m deep is stratification. In the spring and summer deeper reservoirs tend to form two layers; an upper layer warmed by the sun, the epilimnion, with a high population of algae, and a cold lower layer, the hypolimnion, which is not warmed by the sun and being denser does not mix with the upper layer. The lower layer tends to become anoxic, as the oxygen in the water is used by the organisms within it, and reducing (anoxic) conditions develop. As a consequence, concentrations of dissolved iron and manganese are often high in the lower levels of stratified reservoirs. This is not an immediate problem providing it is possible to draw-off water from different levels. Water for treatment is normally drawn from the higher level of the reservoir, where oxidizing conditions exist and dissolved iron and manganese concentrations are low, although algal levels may be high. However as winter approaches and the upper layer cools down there is often a marked deterioration of water quality as thermal stratification breaks down and the reservoir becomes mixed. This can result in poor-quality water in the entire reservoir in the period following destratification.

Reservoirs can be prevented from stratifying by installing mixing systems to mix the hot and cold layers. A common way of doing this is to pump air into the lower part of a reservoir, inducing an upward current that mixes the reservoir. Such systems can ensure a uniform improved water quality in the entire reservoir. Fully mixed reservoirs may also suffer from algal blooms as algae can control their buoyancy to ensure that they stay at a level at which they receive adequate levels of sunshine. Thus destratification may not resolve the problem of algal blooms. However, there is evidence that intermittent operation of destratification measures in a reservoir leads to significantly lower levels of algal mass in a reservoir.[5] This may be related to the difficulty that algae have in adopting to changing environmental conditions, or to the encouragement of grazing by zooplankton.[6]

Algae are a problem that will not go away, although there are measures that can be taken to reduce the problems they cause. It is only in the very long term that the control of nutrient inputs may alleviate the problem. Thus, it is necessary to ensure that water treatment can cope with algae.

FINE SCREENS

Fine screens are normally the first stage of treatment. The screens may be located at the intake or at the treatment works site, with only coarse screens at the intake. Typically they have an effective aperture of approximately 10 mm and remove filamentous algae, waterweed and small debris, and the larger inhabitants of the raw water.

The openings in a fine screen mesh tend to clog very quickly if there is much debris in the water. It is possible to fit the fine mesh in frames and remove these for cleaning by hand. However, except on the smallest works, or where labour is cheap and plentiful, it is normal to have mechanically cleaned screens. These are often one of the many proprietary forms of mechanical screen constructed on the endless band or drum principle and cleaned continuously by water jets which wash the screenings away along channels.

Drum and cup screens both consist of large diameter rotating drums coated with a mesh screen around their perimeter. Water flows from the inside of the drum out through the mesh. Debris is removed on the screen and washed off into a collection trough as it reaches the highest point of the drum. A cup screen allows water only to enter from one side whereas a drum screen permits water to enter from both sides; the principles are illustrated in Fig. 4.1.

For large flows a large diameter screen is required, depending on water levels and mesh size, a 300 000 m³/day cup screen would be of the order of 5 m dia. with a screen width of 1.5 m. Extensive civil engineering works are required for large drum or cup screens. Band screens reduce the size of civil structure required by using flexible sectional bands. The principle of band screens is illustrated in Fig. 4.2. They are particularly useful where there are large variations in water level, which would require a larger diameter drum than would otherwise be needed. A band screen to treat 300 000 m³/day would require a screen around 1 m wide and a minimum immersion of 4–5 m.

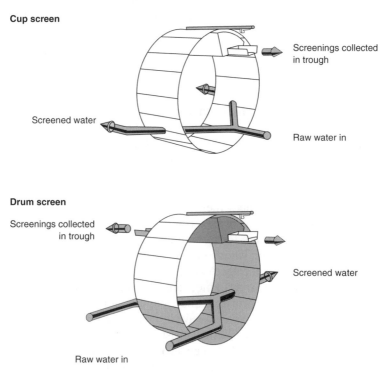

Fig. 4.1. Principle of cup and drum screens (courtesy of Bracket Green)

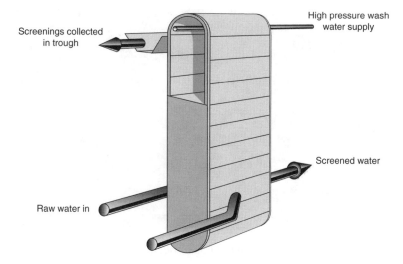

High pressure wash water supply

Screenings collected in trough

Screened water

Raw water in

Fig. 4.2. Principle of band screen (courtesy of Bracket Green)

Drum and band screens are proven and very common but they are relatively complex and expensive. They may also require a significant quantity of water for cleaning; typically this is less than 1% of the total works throughput.

Traditional mechanical screens are complex and two newer types of intake screen are sometimes used on new works in preference. One is a passive submerged screen using a screen derived from the wire wedge screens used in boreholes. The screen itself comprises a cylindrical section manufactured from wedge wire wrapped round a retaining structure. The slot allowing water to enter is tapered, with the narrower width on the outside of the screen to prevent clogging. Figure 4.3 shows a typical arrangement of such a screen. They are cleaned by injecting air into the screen, blasting out debris that accumulates in the wires. Thus such screens must be submerged and have to be placed below the minimum possible water level. They are typically designed to remove particles down to 3 mm or less. Such screens are very simple; requiring relatively minor civil works, and can often be installed at locations where a drum or band screen would require a major structure. An advantage of these screens is that they have minimum impact on fish, only the

Fig. 4.3. Johnson passive screen (USF Johnson)

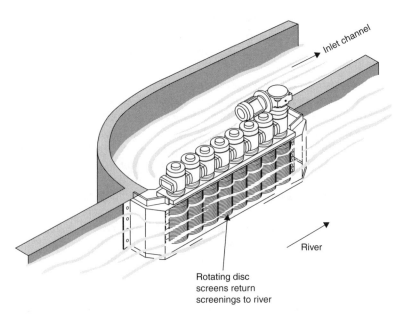

Inlet channel

River

Rotating disc
screens return
screenings to river

Fig. 4.4. Disc screen installation (courtesy of Mono Pumps)

very smallest young fish can pass through the screen. If they are located such that they protrude into a river, making them potentially vulnerable to being struck by boats or large items of debris, adequate protection will be needed.

Another newer form of screen uses rotating discs able to remove particles down to 3 mm or less (Fig. 4.4). These screens are designed to return debris back to the river. They have the advantage of not requiring any cleaning water. Such screens are simple but large installations require significant civil works to allow them to be periodically removed for maintenance.

Key factors to consider in screen design are the variation in river water level and the solids loading in the water being screened. It is common practice to locate river intakes upstream of a weir to provide a minimum water level that is largely independent of river flow. This removes any concern over minimum water level, but rivers can flood and the design must cater for this. Where there is potential for flooding all electrical equipment needs to be above flood level and secure access is required. Table 4.2 compares different types of screen.

PRE-CHLORINATION

Formerly it was common to dose chlorine prior to clarification in order to disinfect, at least partially, water passing through a works and in order to control attached algal and slime growth. Nowadays this is rarely ever practised with surface waters because it maximizes the formation of THMs. For groundwaters with little or no risk of forming THMs pre-chlorination may be used to encourage the oxidation of iron and manganese and to oxidize ammonia. For a typical groundwater containing iron and manganese a chlorine dose of around 0.8 mg of free chlorine per mg of iron or manganese should be dosed as the water arrives at the treatment plant (pH adjustment may also be required). If the objective is to oxidize ammonia a chlorine dose of eight to ten times the NH_3–N concentration is required. (The theoretical quantity is 7.6 times the NH_3–N concentration but in practice this is a minimum.)

AERATION

Aeration can be used to oxygenate the raw water and to release CO_2 and hydrogen sulfide from the water. The rate of absorption or

Table 4.2. Advantages and disadvantages of different screen types

Type of screen	Size of solids removed	Advantages	Disadvantages	Typical uses
Manual bar	Down to around 25 mm	• Simple • Cheap	• Cost of manual operation • May blind if not cleaned • Unsuitable for deep intakes	As very coarse protection to a mechanical screen
Mechanical bar	Down to 12 mm or less	• Relatively cheap	• Arguably more prone to failure than other mechanical screens • Cannot remove small debris • Unsuitable for deep intakes	Not normally used in water treatment
Drum (or cup) or band screen	Down to 1 mm. 5 mm more normal	• Reliable • Drum screens relatively simple • Band screens have less extensive civil works and suitable for deep intakes and for sites with large changes in water level	• Drum screens require more extensive civil works • Drum screens unsuitable for deep applications or for large changes in water level • Band screens more complex • Supply of washwater required • Screenings need to be handled	Both types widely used

		Advantages	Disadvantages	Applications
Disc screen	Down to 2.5 mm	• Simple and reliable • Screenings returned to main flow • No washwater required • Relatively fish friendly	• Need provision for removing entire unit for maintenance • Unsuitable for large changes in water level • Unsuitable for high solids loadings	Reaches of rivers where water level fairly constant
Passive well screens	Down to 2.5 mm	• Very simple • Screenings returned to main flow • No washwater required • No power supply required at screen • Minimum maintenance requirements • Fish friendly • Minimum civil works required • Can be designed to allow screens to be lifted for inspection	• Vulnerable to damage from boats and large debris unless adequately protected • Need air supply for cleaning	River intakes

release of gas by water depends on the concentrations of the gas in the water and the air to which it is exposed, and the surface area across which the gas is transferred. Aeration is also used in treating high levels of iron and manganese. If present they are normally in solution but when oxidized are precipitated and can be removed by filtration. Raising dissolved oxygen levels is an important process to speed oxidation. Cox[7] lists ten different treatment processes that may be used for reducing iron and manganese concentrations; six of these incorporate aeration.

When water is in contact with the atmosphere, gases are absorbed or liberated such that the partial pressure of the gases in the water will move towards the partial pressure of the gases in the atmosphere. If the partial pressure of a gas in the water is higher than the partial pressure in the atmosphere, then the gas will be released from the water; conversely if it is lower, the water will absorb gas. Thus, water with a low dissolved oxygen concentration will absorb oxygen from the air, improving its taste; water containing CO_2 or hydrogen sulfide will tend to lose them, with benefits to the taste and corrosiveness of water. Hydrogen sulfide is essentially non-existent in the atmosphere and it will be almost totally removed in a properly designed aeration plant. However, the removal of carbon dioxide will only be partial, partly because it is present in the atmosphere and partly due to the chemical relationships between CO_2 and bicarbonate in the water being aerated.

Aeration is a cheap and valuable means of controlling tastes and odours due to hydrogen sulfide, of increasing the concentration of oxygen in water, and of reducing the corrosiveness of water. There are, however, limits on its practical effectiveness. Aeration is also used to remove volatile organic chemicals and this is covered in Chapter 11.

Aerators are commonly used if any of the following conditions are present in the raw water:

- hydrogen sulfide (tastes, odours etc.)—to liberate the dissolved gas;
- CO_2 (corrosive tendencies)—to liberate excess gas, raising pH;
- iron and manganese in solution—to lower the carbon dioxide concentration and increase the oxygen content to encourage oxidation of the reduced soluble forms of iron and manganese; and
- low levels of dissolved oxygen—to increase the oxygen content.

Aeration is commonly used on groundwaters containing high levels of dissolved CO_2 or high concentrations of iron or manganese and on water drawn from reservoirs, where the lower-level water may be low in dissolved oxygen if the reservoir is not fully mixed. Water from other surface sources is generally in equilibrium with the atmosphere and there is no need to aerate.

Types of aerator

To ensure aeration proceeds at a useful rate it is necessary to maximize the area of contact between the atmosphere and the water. There is a wide range of aerators used in water treatment; some of the common types are considered below.

Spray aerators—these use nozzles which produce thin jets of water, often directed against metal plates for greater efficiency, giving rise to a fine spray which exposes countless droplets of the water to the atmosphere. Spray aerators are very efficient. The nozzles are commonly of 25–35 mm dia. and discharge about 5–10 l/s. To aerate 10 000 m^3/day about 20 nozzles arranged within an area of 25 m^2 are needed. Care must be taken to shelter the installation against wind, or the fine spray may be blown away from the collecting trays. Louvres set in a surrounding wall often shelter the area where the sprays are located.

Cascade aerators—these depend on the turbulence created in a thin stream of water flowing swiftly down an incline and impinging against fixed obstacles. The surface area of liquid exposed is rather limited and they are not very efficient.

Tray aerators—these typically consist of up to five trays, increasing in size from top to bottom. The water falls from tray to tray through a height of around 0.5 m per tray. The total area of the trays in relation to the flow is generally about 0.1 m^2 per m^3/h. This type of aerator is apt to freeze in cold weather and to encourage the growth of algae and other life in warmer climates. However, it is a simple and cheap method, and is widely used, particularly in the UK.

Slatted tower aerators—these were commonly used in the USA and consist of towers that contain a series of horizontal redwood slats. The water is introduced at the top and it cascades down over the slats. These aerators were generally used for waters high in dissolved iron; the iron tended to oxidize and build up on the wooden slats, from which it would periodically slough off. There is a tank beneath the tower to collect the water, and the oxidized iron settles in the tank and is periodically removed.

Diffused air aerators—consist of tanks in which air is bubbled upwards from diffusers laid on the floor, the diffusers being sufficiently fine and numerous to promote a cloud of small bubbles. This type of aerator is efficient provided fine bubbles are produced and there is adequate water depth. The amount of air used can be regulated according to need. Aeration tanks are commonly about 4 m deep and have a retention time of about 15 min. Advantages of diffused air systems are minimal head loss and a lesser space requirement than other systems. The air blowers should deliver about the same amount of air in any given time as the throughput of water. For a throughput of 10 000 m^3/day an air blower of between 2.5 and 5 kW is required.

Packed tower aerators—these are towers packed with media. Water passes down through the media while air is blown upwards. These are highly efficient for gas transfer and are used for removal of volatile organic chemicals as well as for pre-treatment. They are discussed further in Chapter 11.

Relative effectiveness

Assuming a packed tower be the most effective method of aeration, a very approximate measure of relative effectiveness, relative to the CO_2 removal effected by a diffused air installation, would be:

- sprays 90%;
- trays 70%;
- cascades 50%.

In the UK trays were the most commonly used method of aeration because of their cheapness, simplicity and reasonably high efficiency. They are still appropriate for oxygenation but where the prime aim is to remove CO_2, modern plants generally would use packed tower aerators supplied by a specialist supplier. In some older treatment plants an eductor aeration system is sometimes seen; this passes water through a venturi nozzle, and air is introduced into the water at the venturi throat.

ALGAL TREATMENT

Algae are notorious for causing problems with water treatment. If they are not removed they can lead to taste and odour problems in treated water. If they pass to filtration, they can quickly block up filters, leading to very short running times between cleaning. The first preference is to control their numbers in the raw water, either by reservoir management techniques, or by withdrawing water from reservoirs at levels where algae are only present in low numbers. The concentration of algae is also affected by wind, tending to be higher at the downwind end of a reservoir. If algae cannot be controlled in the raw water, the aim is to remove them as soon as possible in the treatment process.

Strainers can be used to remove algae, either rapid sand filters or microstrainers. If the medium is coarse the former are sometimes known as roughing filters and can be worked at high loadings to take the load off the main filters. Such filters are effective at removing some algal species, but other species require coagulation to be effectively removed.[6]

Microstrainers are similar to drum screens but use a fine mesh to remove small particles. They are capable of excellent performance provided that the water is relatively silt-free. They can effectively remove filamentous or multicellular algae but do not cope as effectively with small unicellular algae. The mesh of a microstrainer has openings of around 20–40 μm for algal removal, and throughput velocities are around 10 m/h through the submerged area. The filter elements are cleaned continuously by jets. The volume of washwater is typically 1–3% of the throughput. The loss of head is dependent on mesh size and loading but is normally small (about 150 mm). They work well for some waters but not others; in favourable cases they will reduce the algal load by 80–90%. In practice microstrainers are little used in new works for algal treatment.

Algae control their buoyancy and thus when living cannot successfully be fully removed by means of settling basins. Thus a prerequisite to removing them by settlement is to kill them. After this they still require to be removed. In the past chlorination was commonly used to kill algae, but as mentioned earlier this is now uncommon. However ozone will kill most algae and this can be used as an alternative to chlorination. Once killed, they can be removed in clarifiers; by DAF; or by filtration.

DAF works well for algal removal; it works best with a stable water quality and is well suited for use with reservoir water, where algae are more likely to be a problem, rather than river water.

PRE-SETTLEMENT BASINS

Pre-settlement basins are not used in the UK, where high silt loads are unlikely to occur for extended periods. However they may be used in countries where there can be long periods of high solids loadings associated with very large seasonal variations in river flow due to monsoons.

Pre-settlement tanks, which are always horizontal flow tanks:

- permit the use of upward-flow basins on rivers formerly considered too turbid;
- improve the performance of existing upward-flow tanks overloaded with high levels of suspended solids;
- may be built later if conditions on the catchment deteriorate (a common occurrence in developing countries where jungle-covered catchment is apt to be cleared for development); and
- can be used with horizontal-flow basins to reduce coagulant use.

Upward-flow basins become increasingly difficult to operate once the dry silt by weight exceeds 1000 mg/l. Where this is likely to happen it is helpful to put in a small, non-chemically-assisted, horizontal-flow basin immediately upstream of the vertical-flow basins to keep the peaks of suspended solids well below 1000 mg/l.

Early pre-settlement basins in the USA were designed largely in accordance with findings of Bull and Darby,[8] on the basis of 3 h nominal detention time and an overflow rate of 1.2–1.5 m/h. Such tanks have been used successfully on waters of up to 20 000 mg/l suspended solids. However, 3 h detention is more than needed, and tanks with a detention time of not more than 1 h have been shown to operate well. Typically effluents with considerably less than 1000 mg/l of suspended solids can safely be expected from such tanks, even though the incoming raw water might have suspended solids of 10 000 mg/l or more.

However, the use of pre-settlement tanks does not necessarily mean that all silty sources can be used. Some tropical rivers have such heavy silt loads during the monsoons that it is simply not practicable to treat water taken directly from the river.

5: Coagulation and flocculation

This chapter looks in some detail at what is often a vital treatment process: coagulation and flocculation. This chapter focuses on the theory and practice of destabilization of fine solids present in water; Chapter 6 considers the types of coagulants and their chemistry.

Coagulation and flocculation is an area of water treatment that developed greatly over the last 20–30 years of the twentieth century. Prior to this, plants often had very simple arrangements for dosing and mixing chemicals. These generally worked well with lightly loaded plants, at the cost of high chemical use. With the development of advanced flat-bottomed clarifiers and DAF much more attention has been paid to the destabilization of water containing fine material and the conditions necessary for the formation of easily removed flocs. Also economic pressures have meant that savings in chemical costs have become more important.

COAGULATION AND FLOCCULATION: INTRODUCTION

It is convenient to think of solids as being present in water in three main forms: suspended particles, colloids, and DS (molecules). Suspended particles may be coarse or fine particles of, for example, sand, rocks, or vegetable matter. They range in size from very large particles down to particles with a typical dimension of $10\,\mu m$. Suspended particles will under quiescent conditions either settle or float. Colloids are very fine particles, typically between $10\,nm$ and $10\,\mu m$. Finally there are DS that are present as individual molecules or as ions. Figure 5.1 shows the size ranges of materials present in water. These size bands are approximate and some sources quote slightly different ranges. The colloid size range includes large organic molecules.

Coarse or fine particles are generally relatively simple to remove by either settlement or filtration. DS cannot be removed by physical treatment save by reverse osmosis (although they may be removed

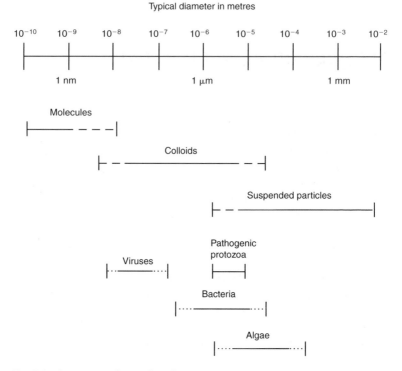

Fig. 5.1. Size range of particles of concern in water treatment

after precipitation). Thus removal of colloids is often the main objective and most difficult aspect of conventional water treatment.

Because the size of colloidal particles is so small they have significantly different characteristics to larger particles. If a 1 m cube of material were to be divided into cubes with a side of 10 nm, the surface area of the material would increase from $6\,m^2$ to $6000\,km^2$. A cubic centimetre of material would have an area of $600\,m^2$. As a result of the large surface area associated with small particles surface phenomena predominate and gravitational effects are unimportant.[1]

Colloids themselves are split into two types: hydrophilic or water-loving colloids, and hydrophobic or water-hating colloids. Hydrophobic colloids are unstable, once the particles aggregate they do not easily reform as colloids. Examples are some clay particles and non-hydrated metal oxides. Other common examples of hydrophobic systems, although of liquid/liquid colloids, are emulsion paint and mayonnaise. Hydrophilic colloids include soap and wallpaper paste. When these materials are mixed with water they form colloidal solutions, which cannot be destabilized.

The reason that colloids stay as such small particles is that the particles have similar negative electrical charges, meaning that electrical forces keep the individual particles separate. The importance of surface electrical charge is associated with the very large specific surface area of the particles.

In order to remove colloids it is necessary to form larger particles that can then be removed by physical treatment. For hydrophobic colloids it is necessary to overcome the forces keeping the particles apart; once this has been done the particles coalesce into larger particles that do not reform into colloids. This process of particle destabilization and formation of larger particles is called coagulation. Flocculation is the process of mixing which results in further collisions between the particles formed by coagulation, and results in the formation of relatively large particles that can be more easily removed.

Hydrophilic colloids cannot be destabilized in the same way as they would simply reform as colloids. They normally have to be removed by chemical precipitation, filtration, or adsorption. However, some hydrophilic colloids can be removed from water by flocculation, specifically those composed of long organic molecules with multiple charges.

COAGULATION AND FLOCCULATION: DEFINITION

These terms are often used interchangeably by water engineers, but this is incorrect. Coagulation is the destabilization and initial coalescing of colloidal particles, specifically hydrophobic colloids. Coagulation occurs extremely quickly. Flocculation is the longer-term process of forming larger particles from the small particles formed by coagulation. Flocculation may also assist in removal of some hydrophilic colloids. Note, however, that sometimes 'coagulation' is used to cover only the destabilization of the particles, and 'flocculation' to cover all agglomeration of particles. These may be more precise definitions but are less commonly used in water engineering.

A coagulant is the chemical that is dosed to cause particles to coagulate. Typically these are metal salts. Polymers may also be used as coagulants but are more often used during flocculation rather than for coagulation. When a polymer is used to strengthen or enlarge flocs formed by coagulation it is normally referred to as a coagulant aid.

TYPES OF DESTABILIZATION

There are generally accepted to be four main methods of destabilizing colloid systems.

Double layer compression—involves the addition of an electrolyte to water to increase the concentrations of ions. This has the effect of decreasing the thickness of the electrical double layer that surrounds each colloidal particle. This allows the particles to move closer to each other, meaning attractive forces have more chance of overcoming the electrical forces that keep them apart. The effectiveness of the coagulant depends on the change in ionic concentration and also exponentially on the charge on the ions added. Thus, ions with a charge of $+3$ are around 1000 times more effective than ions with a charge of $+1$. There are three points to note. Firstly, this method of destabilization only works when ions are present and thus where a metal salt is added which subsequently precipitates as a hydroxide floc, the effect is only present before the insoluble hydroxide is formed. Secondly, the effect is independent of the concentration of colloidal material. Finally, the effect is proportional to the change in ionic concentration.

Charge neutralization—adding ions with a charge opposite to that on the colloidal particles can lead to adsorption of the ions on to the colloidal material and reduction of surface charge. This reduces the electrical forces keeping particles apart and allows easier agglomeration. There are two points to note: the dose needed is proportional to the quantity of colloidal material present, and it may be possible, with some colloids, to overdose, leading to charge reversal on the colloidal matter. Where charge reversal occurs the colloid is not destabilized.

Entrapment in a precipitate—if soluble aluminium or iron salts are added to water at the correct pH value, they will precipitate as hydroxide flocs. If colloids are present then the hydroxide will tend to precipitate using colloid particles as nuclei, forming floc around colloid particles. Once the hydroxide floc has formed it may physically entrap other colloidal particles, particularly during subsequent flocculation. The point to note is that there is often an inverse relationship between the concentration of colloidal material to be removed and the coagulant dose required. This is explained by the concept that at high colloid concentrations the colloidal particles act as nuclei on to which the coagulant precipitates. On the other hand at low colloid concentrations an excess of precipitated coagulant is required to entrap the colloid particles. The optimum coagulation pH value is dependent on the pH/solubility characteristics of the coagulant used.

Particle bridging—large organic molecules with multiple electrical charges are often effective as coagulants. In water treatment such chemicals are normally referred to as anionic or cationic polymers. These are believed to work by bridging between particles. It is interesting to note that both anionic and cationic polymers are often found to be capable of coagulating negatively charged colloid particles. Polymers are also often used during flocculation to aid in particle formation when they are referred to as coagulant aids. Excessive agitation of flocs formed using a polymer coagulant can lead to the flocs breaking up.

COAGULATION AND FLOCCULATION

The aim of coagulation and flocculation is to produce particles of a size that can be removed by settlement, flotation, or filtration. Thus where small particles are formed by coagulation it is normally necessary to have a subsequent flocculation process. Flocculation occurs when particles collide with each other. Flocculation arises from three main processes: Brownian motion, stirring, and differential settling. Because the particles involved in flocculation are much larger than colloid particles the effects of surface charges are much less and do not significantly affect flocculation.

For initial flocculation of particles smaller than 0.5 μm, Brownian motion is the main process. However, as particles increase in size it is necessary to encourage collisions by stirring. The amount of energy used for this needs to be appropriate to the size and strength of the flocs. Too little energy will result in low rates of floc formation but excessive energy input will lead to floc breakage.

ENERGY INPUT TO MIXERS/FLOCCULATORS

The stirring of water creates differences of velocity and therefore velocity gradients. The energy input to a mixer is dissipated by velocity gradients and the unit rate of energy input is proportional to the velocity gradient established. Mixing is commonly either by a baffled channel, a mechanical mixer, or a weir. The velocity gradient in a shearing fluid is denoted by G and this is used to measure the intensity of mixing. A high G denotes violent mixing; a low G denotes gentle mixing.

A general rule is that coagulation requires high-energy mixing. This is because where metal salts effect coagulation by double layer compression and charge neutralization their effectiveness is greatest when the salts are present as ionic complexes, and these complexes only exist for a very short time, of the order of a second or less. For effective economic coagulation by double layer compression and charge neutralization, intense mixing is required to ensure that the metal coagulant is distributed rapidly through the water before insoluble salts are formed. The requirement for violent mixing is less where coagulation is effected by particle bridging and entrapment in a precipitate. This is because floc formation takes a few seconds reducing the need for rapid dispersion of coagulant and high-energy mixing.[2]

Flocculation requires a lower-energy input. High-energy mixing will tend to break up large flocs, which is not at all what is wanted. Ideally the energy used for coagulation and flocculation will be adjusted to the particular process needs. Coagulation will have a high-energy input, with the energy input for flocculation decreasing as the floc size increases. The key design parameters for the design of coagulation and flocculation processes are the intensity of mixing, which the velocity gradient G is used to denote, and the retention time T. Because of the different energy requirements coagulation, or mixing, and flocculation are normally carried out as separate processes.

VELOCITY GRADIENT

The basic equations relating to mixing and flocculation were published some years ago in the USA by Camp and Stein.[3] They showed that the power dissipated per unit volume of a fluid is given by:

$$P/V = \mu G^2 \tag{5.1}$$

where:
     G is the velocity gradient, s^{-1}
     V is the volume of fluid, m^3

P is the power consumption, W

μ is the dynamic viscosity of water, Pa s.

Thus, G is equal to $(P/\mu V)^{0.5}$. This applies either to a mechanical mixer or a flocculator, where the power 'P' is the power transferred from the mixing blades to the water. However, it is also common to dose chemicals into the flow over weirs and to flocculate in baffled or sinuous channels. In this case the power input to the water is determined by the head loss over the weir or in the channel. The power needed to transfer a flow of liquid equal to $Q\,m^3/s$ across a height difference of h is given by:

$$P = \rho g h Q \qquad (5.2)$$

where P is in watts, Q is the flow in m^3/s, ρ is the density of the liquid in kg/m^3, and g is acceleration due to gravity.

Thus for sinuous or baffled flocculation channels:

$$G = (\rho g h/\mu T)^{0.5} \qquad (5.3)$$

where T is the detention time.

Similarly where coagulation is induced by injecting chemicals into the flow over a weir equation (5.3) applies, where 'h' is the fall over the weir and 'T' is the retention time in the chamber into which the weir discharges. However, often the weir is located in a channel. In this case there is no defined chamber and the volume within which the energy is dissipated will have to be estimated by the designer.

GT

The intensity of mixing is one of the key design parameters in mixing and flocculation. The second parameter is the time for which the mixing is maintained. This is easy to calculate, being the volume in the mixing zone divided by the throughput. In practice it is normal to specify both G, the velocity gradient, and GT, the velocity gradient multiplied by the retention time. For mixing, GT has less importance than G, but for flocculation GT is an important parameter. Appendix 1 includes a sample calculation for GT.

TYPES OF RAPID MIXERS (FOR COAGULATION)

The purpose of rapid mixing is to speedily disperse the chemical being dosed throughout the water being treated and where appropriate to agitate the water sufficiently to allow coagulation by double layer compression or charge neutralization. Where destabilization is by particle bridging there is no need for such violent agitation.

It appears clear from much research work that in general the higher the G-value in a rapid mixer the better. High G-values have been shown to minimize the coagulant dose required[4] and to maximize floc size under a particular set of circumstances. It is also clear that most water-treatment plants work well using very low G-values. There are, however, two points that should be noted: chemical usage is minimized by maximizing the effectiveness of coagulation and flocculation, and for more advanced clarification processes, plate settlers and DAF, effective coagulation is essential as retention times in the clarification process are short and the feed water needs to have been properly conditioned.

Mixing devices for metal coagulants should be designed for a high G, of 1000 s^{-1} or higher. For minimum chemical use where destabilization is by charge neutralization or double layer compression, a G of over 5000 s^{-1} is required to minimize chemical dosage. Polymers require a lower value of around $400\text{--}1000 \text{ s}^{-1}$.[5] Detention time in the mixing zone can be very short providing there is effective mixing. Dispersion of simple metal coagulants is assisted by using carrier water to dilute them to a concentration where the density and viscosity of the coagulant solution is near to that of water; however, for polymerized salts (see Chapter 6) this is not recommended as it reduces their effectiveness.

Where pH control is needed to adjust coagulation pH value, the chemical used for adjusting pH should be added prior to dosing coagulant. Chemicals used for pH adjustment need to be properly dispersed but there is no specific requirement for high-intensity mixing.

TYPES OF RAPID MIXERS

There are many forms of rapid mixers.[6] Some are considered below.

Weirs or flumes—it is common to find coagulant dosed above a weir. The mixing takes place in the violently agitated water below the weir. An advantage of this arrangement is that the unit energy input will be largely independent of flow. The main problem is that the coagulant should be distributed as uniformly as possible along the weir. This is actually quite difficult to achieve for what is normally a relatively low-coagulant flow rate. It is extremely common to see poor-coagulant distribution where this method is used. Poor distribution may require more coagulant use, but often in practice this is not a major concern.

Paddle or propeller mixers—comprising a chamber containing a high-speed paddle or a submerged propeller. Where a high *G* is appropriate it is often found that the retention time should be extremely short, of the order of a few seconds, and this can present practical design difficulties. In practice it may be necessary to use a lower *G* and a more realistic retention time. It is important to inject the coagulant into a region where it will be rapidly dispersed into the full flow.

Turbine mixer—comprises a propeller or turbine that imparts a high velocity gradient to the water before it enters the flocculation zone. It is a development of a paddle mixer creating the maximum *G* for a particular power input. Chemicals are injected at the turbine. Figure 5.2 shows a turbine mixer installed in a channel.

Static mixers—this is a short length of pipe containing fixed blades that create violent agitation of flow passing over the blades. They were developed initially by the chemical process industry for intimate mixing of different chemicals. Static mixers used in water treatment typically have a head drop of around 0.05 bar. They have

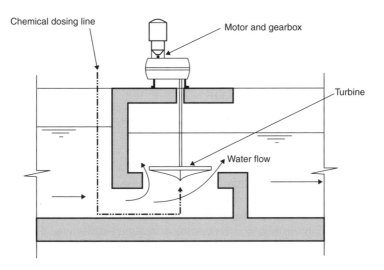

Fig. 5.2. High-intensity mechanical mixer installed in a channel

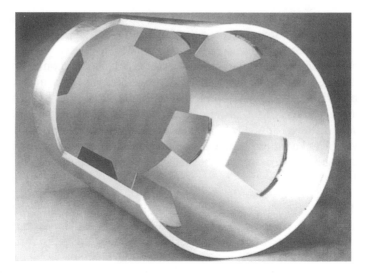

Fig. 5.3. Example of static mixer (courtesy of Chemineer Ltd)

to be installed in a pressure pipeline. Typically they have a length of two to four times pipe diameter. The great advantage of static mixers is that they are compact and highly effective in rapidly dispersing the coagulant. They are arguably less suited to large flow variations as they are optimized to treat a particular flow. Figure 5.3 shows a modern static mixer; some have more complicated blade arrangements than that shown but the principle is the same.

TYPES OF FLOCCULATOR

There are three main forms of flocculators: paddle flocculators, some form of baffled or sinuous channel, and sludge-blanket flocculators.

Paddle flocculators—comprise a series of two or more chambers, each containing a paddle mixer. As the flow passes from chamber to chamber the flocs grow in size, and the G should decrease to prevent the flocs breaking up. It is normal to have variable speed paddles to allow the G to be adjusted depending on water conditions and quality.

Sinuous or baffled channels—are just that. Sinuous channels are fairly common in some African and Asian plants but are unusual in the UK. Flow enters at one end and flows around the channels or baffles. Their advantage is that they are very simple with no moving parts. Their disadvantage is that they are inflexible and cannot be adjusted to take account of differing flow rates or flocculation needs.

Sludge-blanket flocculators—an alternative to flocculators is to pass coagulated water through a bed or blanket of flocs. As the water passes through the blanket smaller particles collide with the flocs to form larger flocs. The concept, developed from simple upward-flow clarifiers, requires a layer of sludge/floc in the bottom of the tank. This has been extensively developed to allow it to be used in flat-bottomed clarifiers (refer to Chapter 8 for more details). Sludge-blanket clarifiers are preceded by a coagulation stage and sometimes by a conventional flocculation stage.

JAR TESTING

An important piece of apparatus in a water-treatment plant is the jar testing equipment. This is used to mimic the coagulation and flocculation process in a laboratory and is used to optimize chemical dosage and coagulation/flocculation pH value. It consists of four or six glass beakers each with a powered paddle which stirs the contents of the beaker. Normally the paddles have a common drive and all rotate at the same adjustable speed. Each of the beakers has the same quantity of raw water added. The paddles are set to rotate at high speed. Then different quantities of coagulant, and possibly acid or alkali to adjust the pH value, are added to each jar. After a short period of intense mixing the paddle speed is reduced to reflect flocculation. After flocculation the jars are observed to assess the size and strength of the flocs formed. If the jars are allowed to settle and the supernatant is filtered through Whatman No. 1 filter paper the quality of the filtered water often approximates to that which would be obtained after filtration through rapid gravity filters.

The jar test is important because it is not possible to predict optimum coagulation conditions based purely on water quality, and also because it is an easy and simple test that plant operators can use for process control.

6: Coagulants and coagulant aids

The previous chapter looked at some of the background theory to coagulation and flocculation; this chapter considers the chemicals used as coagulants and chemicals used to assist in coagulation.

Coagulation theory proposes four methods of destabilization of colloids. The theory suggests that coagulants should have properties to enable the following mechanisms of destabilization.

- *Double layer compression*—compounds forming trivalent cations (positively charged) will be very effective. In practice aluminium and ferric salts are used.
- *Charge neutralization*—compounds forming cations are required. Again aluminium and ferric salts are used.
- *Particle bridging*—simple theory suggests that long cationic molecules would be most appropriate. In practice both anionic and cationic polymers can often be used to destabilize negatively charged colloidal particles.
- *Enmeshment in a precipitate*—compounds forming a hydroxide floc or a carbonate precipitate may be used. Again aluminium and ferric salts are appropriate but other metal salts can be used (for example, calcium hydroxide or magnesium carbonate).

It is clear from the above that aluminium and ferric salts are suitable for three of the four methods of colloid destabilization identified. This has the great attraction that it is not necessary to know the precise nature of the colloids or the destabilization process, and that if the nature of the water changes it will normally be possible to deal with this by adjusting the coagulant dose and the coagulation pH value. This is not to say that all waters can be coagulated with a hefty dose of an aluminium or ferric salt, if the process of coagulation is charge neutralization, overdosing of coagulant can lead to charge reversal and re-suspension of the colloidal material, leading to poorer filtered water quality. However, even if charge reversal occurs, enmeshment may yet rescue the process, although at a high cost in chemical usage.

COAGULATION WITH IRON AND ALUMINIUM SALTS

In the past the common iron and aluminium coagulants were the trivalent compounds: aluminium sulfate, ferric sulfate, and ferric chloride. Table 6.1 summarizes some data relating to these. Suppliers' data sheets should also be used for properties of chemicals used in water treatment.

The addition of a ferric or aluminium coagulant to water sets in motion a complex series of reactions.[1] Initially trivalent ferric or aluminium ions are formed. These then hydrate to form complexes of the metal with water molecules. In a series of further reactions the water molecules are replaced by hydroxide ions giving rise to a further series of complexes. Dependent on pH values and concentrations the iron and aluminium will eventually largely precipitate as a hydroxide floc. The points to note are that many of the intermediate complexes are very effective in double layer compression and in charge neutralization, and they also have a short life. This explains the importance of proper mixing when iron or aluminium salts are used as coagulants. If there is not effective mixing then higher doses of coagulant may be required.

pH control during coagulation with iron and aluminium is most important. The solubility of ferric hydroxide and aluminium hydroxide is lowest at particular pH values for a given water, and above these values concentrations of soluble iron and aluminium higher than the minimum will be carried forward to clarification and filtration. The floc formed by the hydroxides is important in enmeshment and it is normally most effective to maximize its formation during coagulation. However, the salts used for coagulation are strongly acidic, being salts formed from weak bases and strong acids. This makes control of coagulation pH complicated for most waters, and very difficult for un-buffered soft waters.

Coagulation pH

Coagulation pH controls the precipitation of metal hydroxides and also the charges on the intermediate products arising from dosing of metal salts. It is, therefore, most important in ensuring effective and economic coagulation. The usual aim is to coagulate at the pH value where the solubility of the ferric or aluminium hydroxide is at a minimum. This pH value can be theoretically derived from

Table 6.1. Iron and aluminium coagulants

Common name/ chemical name	Chemical formula	Molecular weight	Physical forms	Specific gravity*	Comments
Alum/aluminium sulfate	$Al_2(SO_4)_3 \cdot 14H_2O$	595	Solid granules 8% Al_2O_3, Liquid	Approx. 1.3 (bulk density) 1.32 at 15°C	Normally used in the UK as 7.5% or 8% Al_2O_3 liquid
Ferric sulfate	$Fe_2(SO_4)_3$	400	Liquid[+]	Approx. 1.5	Typically around 40% w/w ferric sulfate content
Ferric chloride	$FeCl_3$	162	Liquid[+]	Approx. 1.45	Typically around 40% w/w ferric chloride content
Polyaluminium chloride	$Al_x(OX)_yCl_z$	High	Liquid	Approx. 1.2	For 10% w/w Al_2O_3
Polyaluminium silicate sulfate	$Al_w(OX)_x(SO_4)_y(SiO_2)_z$	High	Liquid	Approx. 1.3	For 8% w/w Al_2O_3
Polymerized ferric sulfate	$Fe_x(SO_4)_y$	High	Liquid	Approx. 1.6	Approximately 45% w/w ferric sulfate content

*The precise specific gravity will depend on the supplier and the grade supplied. Refer to supplier data sheets.
[+] Also available as solid but this is most unusual for potable-water treatment.

solubility calculations/diagrams for particular water at a given temperature. However, this is impractical in practice and jar testing is often used to determine the optimum pH value for precipitation of hydroxides. Figure 6.1 is a typical simplified solubility diagram for aluminium hydroxide showing how solubility varies with pH value. Aluminium hydroxide can exist in many forms but the diagram only shows the two species that normally determine solubility. Aluminium hydroxide has a minimum solubility at a pH of approximately 7; at lower pH values the solubility increases rapidly and at higher pH values the solubility increases less rapidly. Thus if alum is used the pH value may be set at, or a little above 7 to avoid dissolved aluminium passing to the next treatment stage. However, if charge neutralization is the destabilization process the optimum pH value may be around 5, because at this pH positively charged ions predominate. However, it should be borne in mind that the objective of controlling coagulation pH may not be to minimize

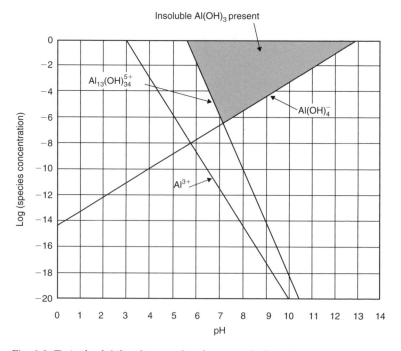

Fig. 6.1. Typical solubility diagram for aluminum hydroxide

carry-over of hydroxide but to remove solids. Thus sometimes the coagulation pH value selected may not be that of minimum solubility, but where this is the case, carry-over of aluminium has always to be considered.

Ferric hydroxide has a minimum solubility over the pH range 7–10, and the increase in solubility at pH values outside this range is less than for alum. Thus ferric coagulants often require a less precise control of pH value.

The ease of pH control depends on the nature of the water. Waters that are high in dissolved salts and alkalinity generally are well-buffered against pH changes arising from dosing coagulants, or strong acids or alkalis to change the coagulation pH. For these waters it is possible to adjust pH using strong acids or alkalis and the final pH value is relatively insensitive to changes in the dosage rates. On the other hand relatively large quantities of pH correction chemicals may be required leading to high chemical costs. Works operators may like such waters as they are easy to control.

Waters that are soft with low DS have very little buffering capacity. Often such waters have organic colour and are acidic. Such waters are poorly buffered and pH-sensitive to strong acids and alkalis, and also to the metal salts used as coagulants, which are strongly acid. It is difficult with such waters to control the pH value within narrow limits and for this reason alum, with a narrow pH for minimum hydroxide solubility, may not be the preferred coagulant. For soft waters the use of metal salts as coagulants will markedly decrease pH.

Aluminium sulfate (alum)

The data in Table 6.1 raises some issues relating to alum. It is possible to easily get confused over strengths of alum solutions and alum dosages. It is common to express the strength of alum as the percentage of aluminium oxide (Al_2O_3). Solid alum has an Al_2O_3 content of approximately between 14% and 17%. Normally in the UK alum is delivered as a solution; typically with a Al_2O_3 content of 7.5% or 8%. The theoretical formula of alum is $Al_2(SO_4)_3 \cdot 18H_2O$. However, the commercial composition of 17% Al_2O_3 alum is $Al_2(SO_4)_3 \cdot 14H_2O$, and this is the formula on which weights and dosage have been calculated in this chapter. An 8% solution has a specific gravity of approximately 1.3. It can then be calculated that to dose 1 mgAl/l to a flow of 1000 m^3/day, 1.63 ml/min of 8% Al$_2$O$_3$ solution is required. In calculating flow rates for other strength solutions due allowance must be made for the specific gravity of the solution. A simplified representation of the reaction of alum with natural alkalinity in water, shown as calcium bicarbonate (but normally measured as CaCO$_3$), is:

$$Al_2(SO_4)_3 \cdot 14H_2O + 3Ca(HCO_3)_2 \rightarrow$$
$$2Al(OH)_3 + 3CaSO_4 + 6CO_2 + 14H_2O \quad (6.1)$$

From this reaction it can be calculated that each gram of hydrated aluminium sulfate requires 0.50 g of alkalinity (as CaCO$_3$) and produces 0.44 g of CO$_2$. In order to buffer the pH it is necessary that the natural alkalinity is in excess of that needed to react with the alum. If there is insufficient then alkalinity will need to be added. This is normally done by dosing lime (normally hydrated lime Ca(OH)$_2$) or sodium carbonate, Na$_2$CO$_3$. For each gram of hydrated aluminium sulfate 0.37 g of Ca(OH)$_2$ or 0.53 g of Na$_2$CO$_3$ is required to provide balancing alkalinity. Note that equation (6.1) is not an accurate reflection of what actually happens and thus calculations based on it are only approximate.

An example of a calculation of alum quantities and sludge production is given in Appendix 1.

Ferric sulfate

In the UK the use of ferric sulfate in water treatment increased in the 1990s. This was for several reasons: the 1989 quality regulations set a limit for aluminium in potable water that could be difficult to attain for some waters if alum was the main coagulant; there was concern over a possible connection between aluminium levels in water and Alzheimer's disease; and the 1989 Camelford incident (where a tanker load of aluminium sulfate was delivered into the treated water tank at a water-treatment works) made all water suppliers wary of using alum. In fact ferric sulfate can be a superior coagulant to aluminium for some waters and it is now more widely used on its merits. It is more pH tolerant than alum giving it advantages in certain processes such as decolorization of waters of low pH value, removal of manganese at high pH value, and the clarification of water of low temporary but high permanent hardness.

The simplified reactions with natural alkalinity are:

$$Fe_2(SO_4)_3 + 3Ca(HCO_3)_2 \rightarrow 2Fe(OH)_3 + 3CaSO_4 + 6CO_2$$
or
$$Fe_2(SO_4)_3 + 3Ca(OH)_2 \rightarrow 2Fe(OH)_3 + 3CaSO_4$$

Ferric sulfate has similar properties to alum and most installations can use either without any particular problems.

Ferric chloride

Ferric chloride solution is more corrosive than alum and has a reputation for being difficult to handle, store and dose. It is normally delivered as a liquid for potable use in the UK, although overseas it is sometimes delivered as a solid. The liquid is little more difficult to store and dose than alum or ferric sulfate: its reputation most likely dates back to the days when it was normally delivered solid. Use of ferric chloride is now common in the UK. Its advantages are similar to those of ferric sulfate. It reacts with natural alkalinity similarly to ferric sulfate.

Polymerized aluminium and iron salts

In the 1960s research into inorganic coagulants led to the development of partially hydrolysed polymeric aluminium salts. These are inorganic polymers with very high molecular weights. The salts are sold under various trade names. The formulation may also include other chemicals to assist in floc formation. Examples are polyaluminium chloride; and polyaluminium silicate sulfate.

These coagulants are more expensive than alum but normally have a number of advantages:

- more effective at low temperatures;
- faster floc formation;
- lower dosage rates;
- savings in pH adjustment chemicals; and
- possibly more effective with algae.

These chemicals can offer significant benefits at plants where there are poor dosing and flocculation facilities; the replacement of alum by a polymeric aluminium salt may greatly improve flocculation and overall treatment efficiency. Some operators have found it effective to use alum when water temperatures are higher, switching to a polymeric salt in the winter when the water is colder. The salts are sold with an Al_2O_3 content similar to alum, making it easy to change coagulants.

Polymerized iron salts have also been developed offering the same advantages as the polymerized aluminium salts. They are used where they offer advantages over aluminium-based coagulants.

Organic polymers

Organic polymers are discussed below in the section on coagulant aids. However, for some waters they are used as the main coagulant. Their particular advantage is that they can be used over a wide pH range. Water-quality standards may limit dosages of polymer dose as noted in Chapter 2.

OTHER COAGULANTS

The coagulants discussed above are those normally encountered in potable-water treatment in the UK. However, a number of other coagulants may be encountered; these may be used in industrial water treatment or overseas. These include:

- ferrous sulfate (also known as copperas) ($FeSO_4 \cdot 7H_2O$) and chlorinated copperas—formed by chlorinating ferrous sulfate;
- sodium aluminate is a compound of sodium oxide and aluminium oxide;
- activated silica—reported to be highly effective as a coagulant aid. It is actually a freshly made sol (a liquid/liquid colloidal suspension) of silicic acid in water, prepared from sodium silicate 'activated' by one of several chemicals.

WHAT COAGULANT?

It is very difficult to state reasons for selecting a particular coagulant. Table 6.2 compares in very simple terms the effectiveness of the three main sorts of coagulants for four different types of waters, using a broad categorization of both waters and coagulants. This table also shows that there are waters that are easy to treat and waters

Table 6.2. Comparison of various coagulants[1]

Type of water	Alum	Ferric salts	Polymer
Type 1: high turbidity, high alkalinity (easiest to coagulate)	Effective over pH range 5–7. No need to add alkalinity or use coagulant aid	Effective over pH range 6–7. No need to add alkalinity or use coagulant aid	Cationic polymers usually very effective. Anionic and non-ionic may also work
Type 2: high turbidity, low alkalinity	Effective over pH range 5–7. May need to add alkalinity to control pH. Coagulant aid not needed	Effective over pH range 6–7. May need to add alkalinity to control pH. Coagulant aid not needed	Cationic polymers usually very effective. Anionic and non-ionic may also work
Type 3: low turbidity, high alkalinity	Relatively high dose needed to form sufficient floc. pH near to 7. Coagulant aid may help	Relatively high dose needed to form sufficient floc. Coagulant aid may help	Will not work well alone due to low turbidity. Adding a clay to increase turbidity may be effective
Type 4: low turbidity, low alkalinity (most difficult to coagulate)	Relatively high dose needed to form sufficient floc. pH near to 7. Alkalinity or clay need to be dosed to produce Type 2 or 3 water	Relatively high dose needed to form sufficient floc. pH near to 7. Alkalinity or clay need to be dosed to produce Type 2 or 3 water	Will not work well alone due to low turbidity. Adding a clay to increase turbidity may be effective

Low turbidity <10 NTU, high turbidity >100 NTU, low alkalinity <50 mg CaCO$_3$, high alkalinity >250 mg CaCO$_3$.

that are difficult. The real problems of coagulant selection arise when water that is difficult to coagulate has to be treated.

Aluminium salts are the most commonly used coagulants, and by adjusting dosage and coagulation pH can be used to treat most waters. Iron salts are similarly flexible.

Other chemicals used in coagulation and flocculation

While hard and fast definitions are difficult there are three main types of chemicals other than primary coagulants used in coagulation and flocculation: coagulant aids, chemicals for pH correction, and weighters.

Polyelectrolytes

The term polyelectrolyte is used as a generic term to describe high molecular organic polymers used in coagulation and flocculation. These are normally synthetic chemicals, such as polyacrylamides or polyamines. Examples of polymers derived from natural substances include sodium alginate or starches. If synthetic, their composition and molecular size can be varied to suit operational requirements. Polyelectrolytes made from polyacrylamide, a linear resin, can be made anionic, cationic or non-ionic, and the strength of the charge and the size of the molecule can be adjusted to suit particular needs. There are health concerns over the use of some polymers in water treatment and the 1998 EU Drinking Water Directive includes limits for some polymers and additionally the UK Government has regulations controlling dosage rates and, for some products, the composition of the polymer.[2]

The high cost of polymers is claimed to be offset by the very small dosage and the saving in pH adjustment chemicals. Dosages are typically between 0.01 mg/l, where used as a coagulant aid, and up to 0.5 mg/l, where used as the main coagulant. In the UK, regulators currently set limits of 0.2 mg/l average and 0.5 mg/l maximum for commonly used polymers. A particular advantage of polymers in some instances is that they can be effective at relatively high pH values on waters containing iron and manganese. This can permit single efficient single-stage removal by filtration of both metals.

Polyelectrolytes are available as powders, beads, or liquids. An everyday example of a polymer is wallpaper paste. As with wallpaper paste, once the polyelectrolyte has been mixed with water, time is needed for the solution to thicken, a process known as ageing. Normally polyelectrolytes are delivered as a powder. The polyelectrolyte is then put into an automatic dissolving system. The polymer powder is placed into a hopper. It is then fed and mixed with water often in an eductor or a mixer into which sprays of polymer powder and water are mixed. The sophisticated mixing devices are needed to ensure proper wetting of the polyelectrolyte with water and to ensure a known strength of mixture. The mixture then passes to an ageing tank, where it is gently mixed to become homogeneous. It is then transferred to a stock or day tank, from where it is dosed. Solution strengths of polymers are low to ensure adequate dispersion, and are typically of the order of 0.5%, or 5 kg of polymer in 1 m^3 of water. Thus relatively large volumes of solution may be required, particularly where polymer is used as the primary coagulant, and automatic polymer make-up equipment is normally used. Some polymers have fairly short shelf life once made up as stock solution and require to be used within a day or two.

However, the more usual use of polyelectrolyte is a coagulant aid, to strengthen weak flocs formed by some waters. Where a metal salt is being used as the primary coagulant, polyelectrolytes often provide a powerful auxiliary bridging and linking action to promote more rapid settlement. It is normally advantageous to put them into the water after the coagulant (i.e. after flocculation has commenced). Polymers may also be dosed prior to filtration as a filtration aid and are also important in sludge treatment.

Lime

Lime is widely used in water treatment, for pH correction. The word lime is loosely used to cover both CaO (quicklime) and Ca(OH)$_2$ (hydrated or slaked lime). Quicklime is slaked prior to use. This is done by mixing controlled proportions of quick lime and water. This is done using continuous slaking equipment, possibly feeding to a slurry tank. Key parameters in slaking are temperature rise and slaking time. The equipment is relatively complex and the process has a bad reputation for being unreliable. For this reason it is uncommon to find water-treatment plants using quicklime; slaked lime is more usual.

Lime is normally delivered in bulk as a powder. Quicklime is then slaked: slaked lime is made into a slurry. Lime is dosed as a slurry typically containing around 5% CaO. The concentration of the slurry is important to minimize problems with blocked feed lines due to deposition. Lime dosing plants do not have a good reputation for reliability and standby equipment and dosing lines are always provided, together with easy access for clearing blockages.

Weighters

Where enmeshment in a precipitate is the main coagulant process, theory suggests that the dose of coagulant required will tend to be inversely proportional to the concentration of solids that will be removed by coagulation. For waters having low concentrations of colloidal matter it is sometimes found that large doses of coagulant are needed to ensure adequate removal of colloids. For such waters it can be effective and economical to add fine material to increase the solids concentration in the coagulation/flocculation processes.

Even where double layer compression or charge neutralization is the primary method of destabilization enmeshment will normally be important in the flocculation process and the addition of a weighter may assist.

Bentonite and Fuller's Earth can be used as weighters. They have been used on some soft-coloured waters, notably in the north of England and Scotland. However, it is unusual to find them being used, with water companies preferring to limit the number of chemicals used if at all possible.

CONTROL OF COAGULATION AND FLOCCULATION

Given that coagulation and flocculation are now seen as important treatment processes in their own right, it is clearly important to control them properly. This is important not only to ensure optimal treated-water quality but also to minimize chemical dosing costs. All control methods depend on flow-proportioned dosing of chemicals, with the dosage determined by one of a number of alternative methods.

In the past the dosages were defined by undertaking regular jar tests to confirm optimum coagulation pH, coagulant dose, and flocculation conditions. Although jar testing does not accurately reflect what is actually happening in the process, the coagulation/flocculation requirements predicted from jar testing are normally in practice a good indicator of the actual optimum. However, jar testing requires regular operator input and as a result other, automated, methods of determining chemical dosages are becoming more common.

Coagulation is sometimes controlled by an instrument called a streaming current detector (SCD). This is used where destabilization is predominantly by double layer compression or charge neutralization. The SCD takes a sample of coagulated water and forces it through a small cell containing a reciprocating piston. By measuring the current generated as the particles in the water pass the piston the charge on the particles can be deduced. It is then possible to control coagulant dose to ensure that the optimum charge for coagulation is maintained on the particles. Although the theory of SCDs appears attractive there are often practical difficulties and they are not widely used in the UK.

A more empirical approach is to control coagulation using historical data on raw-water quality, coagulant dose and pH value, and treated-water quality. This approach requires a large amount of data to enable the relationships between raw-water quality, coagulant dose and pH value, and treated-water quality to be established. Where this data exists the coagulation control can be either manual or automatic, using suitable water-quality instrumentation.

Alternatively coagulant dosages can be controlled by continuously monitoring water-quality downstream of coagulation and using these readings for control. Typically pH, turbidity; colour; and residual coagulant would be measured downstream, possibly at more than one point in the process, and used to control coagulation.

CHEMICAL STORAGE AND DOSING

Apart from lime and polyelectrolyte, most chemicals in the UK are delivered as a liquid. This allows easy delivery and storage with uniform chemical strength, and avoids the process of making a solution. Polyelectrolyte for water treatment is normally delivered as a solid although some polymers are available as a liquid. The drawback to using liquid chemicals is that much of what is delivered is water. Where delivery distances are short this may be economical, but delivering water long distances by road to the water-treatment works clearly may not be economic where a solid chemical could be delivered. In particular in many parts of the world alum is delivered as a solid—as a powder, in small lumps, or in large blocks of over 100 kg. It then has to be dissolved before being dosed—although there is at least one treatment works in Asia where alum is 'dosed' by placing blocks in the hydraulic jump downstream of a standing wave flume.

Table 6.3 lists some of the chemicals normally used in coagulation and flocculation, either as coagulants or for pH control, and gives some details on storage and dosing. All of the chemicals listed need to be treated with care, and some are extremely dangerous. In designing or specifying systems for these chemicals it is essential to consider health and safety aspects and how to safely cater for failures of plant. Liquid storage areas need to be bonded and drained to a storage/neutralization tank. Different types of flanges may be used to ensure tanker deliveries cannot go to the incorrect tank. Emergency showers are needed for most of the chemicals listed.

Nowadays most liquid chemicals are stored in above-ground GRP tanks. However, not all GRP resins are suitable for all chemicals and specialist advice is needed in specifying chemical storage tanks. Polyvinyl chloride (PVC) tanks may also be used for most of the above chemicals, but again advice should be sought from manufacturers.

Chemical dosing has to be precisely controlled for trouble-free water treatment. This requires the use of positive displacement pumps that dose a constant quantity of chemical for each pump cycle. Until recently chemical dosing has typically been done using variable stroke diaphragm pumps powered by variable speed motors. The quantity of chemical dosed per unit volume of water being dosed (in other words the dose in mg/l) was controlled by varying the stroke of the pump. This could either be set manually or by an instrumentation and control system. Varying the pump speed, using the signal from a flowmeter, catered for variations in flow. Systems of this sort have been used successfully for many years and continue to be used. The design of chemical dosing pumps takes into account the chemicals being dosed, and the nature of the water being dosed. The materials in the pump head are dependent on the chemical being dosed. Where chemicals are being dosed to a potable-water system then the pump has to be designed to prevent harmful substances from the pump entering the water.

The weakness with such systems is the chemical pump itself. While traditional dosing pumps are extremely reliable they commonly have two problems. If there is a blockage in a chemical line they are unable to operate at the higher pressures needed to clear it, and they commonly use a diaphragm which requires regular replacement, or which would eventually fail. The advent of sophisticated computer-controlled chemical dosing has allowed the use of other forms of positive displacement pumps which are controlled only by varying pumping speed, with the control system adjusting pump speed to take account of both flow and dose. Small progressive cavity pumps are now increasingly used for dosing of chemicals. They are particularly useful for lime dosing where lines are prone to blocking. The use of positive displacement pumps does not completely overcome problems of blocked chemical lines but it does reduce their impact on plant operation.

Examples of calculation of chemical dosages are given in Appendix 1.

Table 6.3. Storage requirements for various chemicals used for coagulation

Chemical	Delivery form	Storage requirements	Dosing requirements	Comments
Aluminium sulfate (alum)	Liquid	Normally stored in lined concrete tanks or glass-reinforced plastic (GRP) storage tanks	Diluted by carrier water after dosing pumps*	Strongly acid
Aluminium sulfate (alum)	Solid-small blocks (kibbled)	Normally delivered directly into concrete saturators	Diluted by carrier water after dosing pumps*	Saturated solution should be transferred to a day tank and diluted to a standard strength for ease of control. Strongly acid
Aluminium sulfate (alum)	Solid-large blocks	Dry storage	As for liquid	Blocks placed in saturators as for kibbled material
Polymerized aluminium salts	Liquid	Normally stored in lined concrete tanks or GRP storage tanks	May be diluted with carrier water*	Acid
Ferric salts	Liquid	Normally stored in lined concrete tanks or GRP storage tanks	Diluted by carrier water after dosing pumps*	Strongly acid
Hydrated lime	Powder	Silo	Prone to blockages— carrier water and duplicate dosing lines normal	Batches of solution made up using precise quantities of lime and water. Strongly alkaline
Sodium hydroxide	Liquid	Normally stored in lined concrete tanks or GRP storage tanks		Strongly alkaline

| Polymers | Powder-bags | Dry storage | Carrier water used after dosing pumps | Batches of solution made up using precise quantities of polymer and water. Solution requires ageing prior to use. Solution very slippery on floor |
| Acids | Liquid | GRP or lined steel tanks | Carrier water sometimes used after dosing pumps | Carefully designed storage essential |

*An excess of carrier water will affect the coagulation process, particularly with the polymerized products for which the supplier's instructions should be followed.

7: Theory and principles of clarification

This chapter considers some of the principles that apply to clarification, covering settlement under quiescent conditions, shallow-depth settlement, upward-flow tanks and DAF. It also covers sludge-blanket (or upward-flow solids contact) clarifiers which combine flocculation, in a layer of flocs, with upward-flow clarification.

THEORY OF SETTLEMENT

When silt-laden water is admitted to the still conditions of a sedimentation basin and its velocity falls to near zero, its capacity to transport solids disappears, and suspended solids will begin to settle or rise, depending on whether their density is greater or less than water. Assuming that no forces other than gravity are involved, a particle with a density greater than water will settle. The particle, assuming it to be spherical, will accelerate until it reaches its terminal velocity, given by:

$$V_s = [4g(\rho_1 - \rho)D/3C_D\rho]^{0.5}. \qquad (7.1)$$

Where:
D = diameter
ρ_1 = density of sphere
ρ = density of water
g = gravitational acceleration
V_s = settling velocity
C_D = drag coefficient.

Equation (7.1) is derived from the definition of coefficient of drag, which is the ratio of the actual drag force to the dynamic drag force:[1]

$$C_D = F/(0.5\rho V^2 A).$$

Where:

F = actual drag force

V = relative velocity

A = projected area of the moving body.

Where gravitational forces and viscous forces act on bodies, the relative size of the two forces is defined by the Reynolds number, where:

$$Re = \rho VD/\eta. \tag{7.2}$$

Where:

Re = Reynolds number

V = velocity of the body

η = dynamic viscosity.

For Reynolds numbers below 500, flow is predominantly laminar, whereas for Reynolds numbers greater than 2000 flow is predominantly turbulent.

In 1946 Camp[2] collected and published the results of work into the relationship between the Reynolds number and C_D for spheres. The relationship is shown in simplified form in Fig. 7.1.

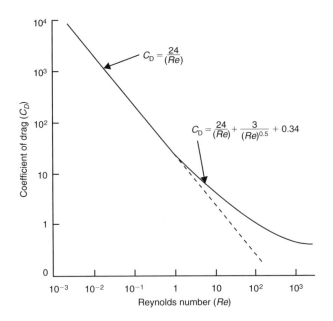

Fig. 7.1. Relationship between C_D and Re

For the straight part of the curve, which represents laminar flow

$$C_D = 24/Re.$$

Substituting for C_D in equation (7.1) leads to:

$$V = g(\rho_1 - \rho)D^2/18\eta. \qquad (7.3)$$

Which is Stokes' law. For spheres with the same density; providing the Reynolds number is low, the unhindered settling velocity is proportional to the diameter of the sphere squared. As the Reynolds number increases to more than approximately 2000, C_D becomes equal to approximately 0.4: thus the settling velocity is given by:

$$V_s [3.3g(\rho_1 - \rho)D/\rho]^{0.5}. \qquad (7.4)$$

Using formulae (7.3) and (7.4) it is possible to calculate settling velocities for low and high Reynolds numbers. Some key properties of water are given in Table 7.1. The kinematic viscosity of a fluid is equal to the dynamic viscosity divided by the density.

Settling velocity is inversely related to kinematic viscosity, which decreases with rising temperature (Table 7.1). It therefore follows that higher water temperatures decrease the drag coefficient and thus increase the rate of settlement. Similar particles could take twice as long to settle in the winter as in the summer and settlement basins need to be sized accordingly.

Table 7.1. Variation of viscosity of water with temperature

Temperature (°C)	Density of water (kg/m^3)	Dynamic viscosity η (10^{-3} Pa s)	Kinematic viscosity v (10^{-6} m^2/s)
0	999.87	1.7921	1.7923
4	1000	1.5676	1.5676
10	999.37	1.3097	1.3101
20	998.23	1.0087	1.0105
30	995.68	0.8004	0.8039

SETTLEMENT IN HORIZONTAL-FLOW BASINS

For simple, unaided settlement consider a rectangular horizontal-flow basin as portrayed in Fig. 7.2. It is clear that a particle settling at v m/s and being carried horizontally by water flowing at velocity v_s m/s would follow the inclined path AB, and by comparing similar triangles the particle would just reach the bottom when:

$$v_s/v = D/L.$$

If

  Q is rate of flow, m^3/s
  W is width of basin, m
  A is area ($=WL$), m^2

then

$$Q = vW/D$$

and thus

$$v_s = Q/A.$$

The quantity Q/A has the units of m^3/s m^2 or m/s, and is known as the overflow rate or surface-loading rate. It is equal to the theoretical settling velocity of a particle that would just settle out in the settling volume represented in Fig. 7.2. It is generally stated in metres per hour or, sometimes, in metres per day.

The importance of surface area in the simple theory of settlement is clear and logically it would appear that the depth of the basin and

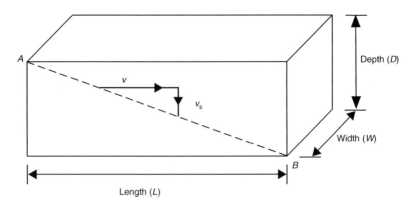

Fig. 7.2. Theoretical settlement in a horizontal-flow settlement tank

therefore the liquid retention time is of little significance. This theoretical conclusion is the basis for tube and plate settlers. However, although the theory is of great importance, there are other factors involved and in practice simple theory does not apply in full.

Laboratory experiments

Much research has been done on the settling velocities of discrete particles and typical figures for the settling velocities of selected particles are given in Table 7.2. These are indicative of the general range of settling velocities of the particles to be removed by settlement; but the velocities are difficult to apply directly. Where a design is to be based on settling velocities, it is necessary to determine actual values by laboratory tests. The figures from laboratory testing should only be applied with caution because the particle sizes in any given suspension are very variable, and the particles rarely remain discrete. There is inevitably interaction between the particles of different sizes, particularly after coagulation, with larger particles falling faster than smaller ones and tending to collide and combine with smaller particles. The particle size distribution of a particular water will vary over the course of a year. The settling velocities in Table 7.2 are significantly less than the values obtained by calculation—the

Table 7.2. Settling velocities of particles in water

Diameter of particle (mm)	Settling velocity (mm/s)		
	Sand, specific gravity 2.65		Alum floc, specific gravity 1.05 at temperature 10°C[5]
	Novotny et al.[3]	AWWA[4]	
1	140	100	0.7*
0.5	72	51†	
0.1	6.7	8	
0.05	1.7	2.9	
0.01	0.08	0.154	
0.005	0.01		

*Calcium softening precipitates with a specific gravity of 1.2 settle approximately three times as fast as alum floc.
†Value interpolated from data.

theory relates to spheres rather than the irregular-shaped objects found in real life.

Under typical waterworks conditions, with flocculation and agglomeration taking place, the simplifying assumption of there being discrete particles does not apply and it is easy to demonstrate that retention time is also of importance, and therefore settlement is not independent of basin depth. One such experiment makes use of a cylinder, equal in depth to the proposed basin, with draw-off points at different levels, as shown in Fig. 7.3. The turbidity of samples from each of the draw-off points A, B, C and D can be measured at increasing time intervals as the water clears slowly from the top. The practical importance of this test is that in a horizontal-flow basin the average turbidity at the outlet weir reflects that of the entire vertical cross-section of the basin and the effectiveness of the basin after any

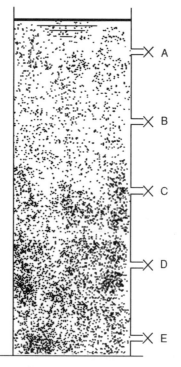

Fig. 7.3. Settling rate experiment

Table 7.3. Results of settlement experiment

Elapsed time (min)	Concentration ratio sampling point						Removal ratio
	A	B	C	D	E	Average	
0	1.0	1.0	1.0	1.0	1.0	1.0	0.0
15	0.65	0.96	0.98	0.99	1.00	0.92	0.08
30	0.23	0.81	0.94	0.97	0.98	0.79	0.21
45	0.10	0.62	0.85	0.93	0.96	0.69	0.31
60	0.05	0.45	0.72	0.86	0.92	0.60	0.40
90	0.03	0.23	0.52	0.70	0.83	0.46	0.54
180	0.01	0.06	0.16	0.32	0.46	0.20	0.80

given time is represented by the average of the turbidity readings at all the outlets.

In a widely quoted experiment carried out by Camp[2] the results in Table 7.3 were noted. The figures show the variation of concentration ratio, defined as the measured suspended solids concentration divided by the initial concentration, with time and depth under quiescent settling conditions. The figures are for discrete particles. The last two lines are the averages of the concentration ratios and, derived from these figures, the removal ratios. The figures show that removal is not independent of depth and is influenced by both overflow rate and retention time.

In this experiment it can be seen that shallower basins would achieve better clarification in a given time. For instance, after 90 min retention the average concentration ratio of silt from points A and B would be 0.13, as opposed to 0.46 for a basin of full depth. The removal ratio in the shallower basin would be 0.87 as compared with 0.54, and the logical conclusion is that for the conditions of this experiment a shallower basin would be better than a deep one.

SETTLEMENT IN UPWARD-FLOW TANKS

Settlement in the upper regions of an upward-flow basin is controlled by making the area (A) of the tank sufficiently big so that $v < v_s$, where v is the upward velocity of the water ($= Q/A$) and v_s is

Table 7.4. Design upward water velocities for upward-flow tanks

Source of water	Velocity (m/h)
River supply, normally coagulated	1.5
River supply, with coagulant aids	3–5
River supply, with coagulant aids and floc improvement by pulsing	6
Softening plants	4.2

the settling velocity of the particle that it is desired to remove. When this condition is achieved the particle will be settling through the rising water and clarification must result, with all particles with a settling velocity greater than v being removed.

In practice the upward velocity of the water is kept down to about half the settling velocity of the floc particles. The normal velocity of settlement of well-formed floc is about 3 m/h.[5] If coagulant aids are used this may become 6–10 m/h, and where the floc is consolidated by pulsing or other devices it may be even higher. In a softening plant the settling velocity of the particles of calcium carbonate is about 8 m/h. Applying the commonly used factor of 0.5, many upward-flow basins are provided with surface areas designed to limit the upward velocity of the water to the values shown in Table 7.4. Chapter 8 provides more details of typical loading rates for different clarification processes.

As in the case of the horizontal-flow basins it would appear in theory that the depth of the tank is not particularly significant and that surface area is the only important factor. However, this conclusion is not correct. Depth is important for a number of reasons and the vast majority of upward-flow basins are fairly deep, typically 3–6 m.

SLUDGE-BLANKET CLARIFIERS (UPFLOW SOLIDS CONTACT CLARIFIERS)

These clarifiers were commonly known as sludge-blanket clarifiers but are also referred to as floc blanket or (upflow) solids contact clarifiers. They are upward-flow clarifiers and their efficiency depends on highly efficient flocculation resulting from passing the feed water through a layer of fluidized flocs. They have been developed from hopper-bottomed upward-flow clarifiers. These were desludged at regular intervals, leading to a build-up of settled sludge in the hopper prior to desludging. It was found that the tanks performed better than was expected and the reason was that as the water entered the tanks it flowed upwards through a layer of sludge. This led to flocculation, with collisions between small flocs in the feed water and the larger flocs in the sludge.

The first sludge-blanket clarifiers were upward-flow clarifiers in which a layer of 'sludge' was maintained in the body of the tank, as opposed to operating the tank to minimize solids retention. However, such tanks are very deep and expensive to construct. Modern sludge-blanket clarifiers are flat bottomed and are much more complex. To work properly they depend on accurate distribution of feed water across the entire tank and precise control of the 'blanket'. The sludge blanket consists of a fluidized layer of flocs, which are settling at a velocity approximately equal to the upflow velocity.

The principles applied to sludge-blanket clarifiers are relatively straightforward. However, the practical application of the principles has led to a wide range of proprietary designs, some of which are extremely complicated. The Pulsator tank is a development of sludge-blanket clarifier whereby a pulse of pressure is regularly applied to the feed water in order to encourage flocculation and to maintain the blanket in a uniform suspension. This is done by the continuous sequential application and release of a small negative pressure to the feed water. Chapter 8 illustrates some of the possible layouts of sludge-blanket clarifiers.

THE MERIT OF ADEQUATE DEPTH

Except for shallow high-rate clarifiers all basins have appreciable depth, typically 3 m or more. Erroneously or not, it is well-established practice to relate basin capacity not only to surface loading but also to the hourly rate of throughput and to classify basins in terms of hours of nominal retention. A capacity of 3–6 h is typical of many horizontal-flow basins but vertical-flow basins are invariably much smaller.

The reasons for providing adequate depth to basins are as follows.

- The theory of settlement is based on the concept of the discrete particle. A discrete particle by definition remains separate and unchanged in volume whereas, in practice particles grow in size by agglomeration with other particles. The bigger the particle becomes the quicker it settles; and the greater the distance through which it can fall the more of its smaller neighbours it adheres to and removes from suspension. Therefore depth (i.e. falling distance) has value.

- Additional depth may be required for the accumulation of sludge. If there is periodic removal of sludge some surplus tank capacity is required for sludge storage.
- The theory of settlement assumes gentle non-turbulent flow, but it is clear that, in very shallow horizontal-flow basins, velocities would tend to increase to a point where excessive velocity and turbulence could cause re-suspension of fine particles and interfere with the whole settlement process. Because $v = Q/$(cross-sectional area of the basin) and the area is equal to the width times the depth, the depth plays a part in keeping v within reasonable limits (i.e. 0.3–0.9 m/min). Above velocities of this order more of the finer particles would remain in suspension.
- In upward-flow sludge-blanket clarifiers, water enters at the bottom and in its passage upwards has to flow through a zone of settling sludge. This action depends on the presence of sludge and is more efficient and easier to control in fairly deep tanks.

SHALLOW-DEPTH (HIGH-RATE) SEDIMENTATION

It has taken many years to realize the undoubted advantages inherent in shallow-depth sedimentation and for tube or plate settlers to become generally accepted. The principle of making settling basins as shallow as possible was first stated by Hazen in 1904. In the 1950s, Camp was proposing settling basins with depths of only 150 mm and detention times of 10 min. Many early attempts actually to do this were unsuccessful because of the unstable hydraulic conditions in the shallow basins and the difficulty of removing sludge. Since the 1960s the use of laminar-flow tilted tube or plate settlers has developed and their use is now commonplace.

The following deals with counter-current systems, i.e. systems where the flow of the water is upwards and the flow of the sludge is downwards. This is not the only arrangement possible and, although rarely encountered, both co-current and cross-current arrangements are possible. Similar principles to those set out below apply to these other layouts. Both plate and tube settlers use the principles and concepts of quiescent settling outlined above. They both require laminar flow to operate. This requires a Reynolds number of less than 800.[6]

PLATES

Under laminar-flow conditions there will be a parabolic velocity gradient for the velocity parallel to the plates, with the velocity being zero at the plates and a maximum midway between the plates as shown in Fig. 7.4. The average velocity compared to the maximum velocity will depend on the geometry of the settler; for circular tubes the mean velocity is half the maximum velocity whereas for plates it will be two-thirds of the maximum velocity.

The settling velocity v_s of the particle theoretically removed in a plate settler is considered by Yao.[7] A simplified approach to arriving at his conclusions can be derived from a consideration of Fig. 7.5. This indicates plates with a settling length of L_s, a plate spacing of d, and an angle to the horizontal of θ. The path of a particle settling at a uniform velocity will be complex because of the effect of the parabolic velocity gradient shown in Fig. 7.4. The actual path will take the form of the curve represented by EF. However this can be approximated to by the straight line BD, where BD represents the path (simplified) of the particle with the lowest settling velocity. It can be seen from inspection that:

$$v_s/v = BA/AD \quad \text{and} \quad v_s = v \times BA/AD$$

but $BA = d/\cos\theta$ and $AD = L_s + d/\tan\theta$.

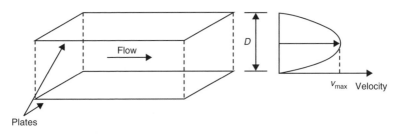

Fig. 7.4. Velocity distribution in a plate settler

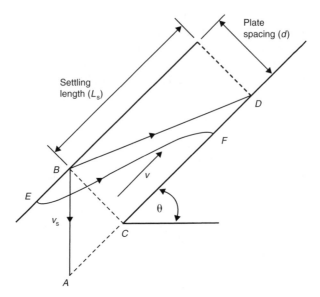

Fig. 7.5. Theory of plate settlers

Thus:

$$v_s = v \times d/\cos\theta \times 1/(L_s + d/\tan\theta).$$

Thus:

$$v_s = vd/(d\sin\theta + L_s\cos\theta).$$

However, this assumes uniform laminar flow which is only found in plate settlers or shallow trays. Thus a correction factor k is introduced. The values of k for circular, square, and parallel plate settlers are respectively 4/3, 11/8, and 1. Thus:

$$v_s = kvd/(d\sin\theta + L_s\cos\theta).$$

It is apparent that the average velocity parallel to tubes or plates inclined at an angle of θ will be given by:

$$v = U/\sin\theta.$$

Where:
   U is equal to the surface-loading rate, or vertical velocity in the tank
   v is the average velocity along the tubes/plates.

Table 7.5. *Decrease in V$_s$ as % of overall surface loading for differing lamella plate spacing and lengths*

Effective length of plate (m)	Distance between plates (mm)			
	100	80	60	40
1	19.7	16.2	12.6	8.6
1.5	13.8	11.3	8.6	5.9
2	10.6	8.6	6.6	4.5
2.5	8.6	7.0	5.3	3.6
3	7.3	5.9	4.5	3.0

Notes:
1. Assumes plates at 60° to horizontal.
2. No allowance has been made for correction coefficient.
3. No allowance made for edge affects.

Table 7.5 shows the effect of plate spacing and effective settling length on v_s. The figures in the body of the table compare the theoretical actual settling velocities to the average surface loading of a tank for plates at an angle of 60°, calculated as percentages. Thus for a lamella plate installation with 80 mm between plates and a settling length of 2.5 m, angled at 60° to the horizontal, the theoretical settling of the slowest particle removed would be 7.0% of the overall surface loading on the tank, allowing a reduction in tank area of 93% to achieve the same degree of particle removal.

The figures in Table 7.5 indicate the importance of both length and plate spacing on the performance of lamella plate systems. However, they need to be treated with a degree of caution. Velocity through the clarifiers is an important factor in the design of highly loaded plate or tube settlers. This needs to be low enough to ensure laminar flow, as discussed above. However, laminar flow is easily achieved with the plate spacings and flow velocities normally encountered.

The figures in Table 7.5 make no allowance for edge effects and thus relate only to very large installations. In practice the additional space required to accommodate the angle and length of the plates means that the figures overstate the theoretical benefits possible. In addition a correction coefficient needs to be applied to allow for turbulence in the inlet and outlet zones, for the effects of sludge movement and removal, and for other factors that will lead to actual performance being worse than predicted. However, even if in practice the performance is only half of that indicated above it still represents a significant improvement over conventional clarifiers. In spite of the marked inclination of the tubes it is important to remember that the settling action is governed by horizontal-flow principles which are in no way different from those described in the early part of this chapter.

There are some practical difficulties associated with lamella clarifiers. Such clarifiers only have very short retention times, often of the order of 15 min within the lamella zone. Such short times are very unforgiving in the event of any upsets to the process or changes in raw-water quality. Thus where such clarifiers are used, the process control will need to be more sophisticated than that required for a plant using conventional clarifiers.

The angle of tilt is critical: too little, and the sludge fails to slide down the plates; too much, and the tube acts on the upward-flow principle, with the benefits due to shallow-depth horizontal-flow action being lost. The material used to manufacture the plates or tubes is of great importance. It needs to be a low-friction material to ensure proper sludge removal. If plastic is used it needs to be resistant to sunlight, it also ideally needs to have good resistance to algal growth, which in practice means that it should be translucent. Counter-current settlers are normally at an angle of around 60°, co-current units can have lower angles (typically 30°) as water movement encourages the sludge to slip down. Co-current units have an arrangement whereby the clarified water is withdrawn at the bottom of the plates prior to the point where the sludge leaves the plates; this is necessary to prevent the sludge being mixed with the clarified water.

A particularly worthwhile benefit of high-rate settlers lies in the ease with which packs can often be inserted into existing basins of conventional design, thus increasing their output capacity without requiring additional land or civil structures. When used in this way the improvements in plant performance will often be very marked as the loading on the modified units will normally be low.

DISSOLVED AIR FLOTATION

It can be seen from equations (7.1), (7.3) and (7.4) that the 'settling' velocity will be positive when $\rho_1 > \rho$ and will be negative when $\rho_1 < \rho$. In the first instance the particles will settle, and in the second instance the particles will rise. Certain suspended solids like algae have a naturally low density and tend to float. If the net density of particles can be reduced to less than that of water by attaching air bubbles to them, then these particles will rise.

DAF involves mixing a flow of water supersaturated with air with a flocculated suspension and introducing the mixed flows into the bottom of a tank. The air comes out of solution, either using the flocs as nuclei onto which the air bubbles form or else becoming attached to flocs as a result of collisions. Once a particle has a net density less dense than water due to attached air, it starts to rise. As it rises the air bubble grows as the pressure decreases and the particle will therefore accelerate upwards. The 'sludge' accumulates as a foam on the top of the tank and is either scraped or flushed off the top. Clarified water is taken out of the tank at low level.

An important parameter in DAF is the ratio of air to solids. This is determined experimentally but it is useful to understand the background theory. Consider a solid particle with an air bubble attached. The density of the particle and air bubble combined is given by:

$$\rho = (V_a \rho_a + V_s \rho_s)/(V_a + V_s).$$

Where:

ρ = average density of solid particle with air bubble attached

V_a = volume of air bubble

ρ_a = density of air

V_s = volume of solid particle

P_s = density of solid particle.

If the masses of air (a) and solid (s) are substituted in the above equation, it can be transformed to:

$$\rho = (1 + a/s)/(1/\rho_s + (a/s)/\rho_a)$$

and then to:

$$a/s = (1 - \rho/\rho_s)/(\rho/\rho_a - 1).$$

For a particle to rise, the density of the particle with air attached must be less than the density of the liquid and the above equation therefore represents the minimum air to solids ratio required. The air density used to calculate the air required should be based on the pressure at the bottom of the tank, and not on atmospheric pressure. Thus the quantity of air required theoretically depends on the density and concentration of solids in the water and the depth of the tank.

In practice the required air quantity will be determined from on-site testing and will be an adjustable parameter in the full size plant.

A typical DAF unit is normally rectangular with a water depth of 2–3 m. Loading rates are normally up to around $10 \, m^3/m^2h$, but can be higher than this. Retention time is around 20 min. The recycle rate is around 6–15% of the flow to be treated and the air requirement is around 6–10 g/m^3 of water treated.[8] The air is compressed to between 4 and 6 bar, depending on the air saturation system employed. The source of the water to be supersaturated is normally the clarified water, in potable-water treatment this is usually of acceptable quality. DAF is always preceded by coagulation and flocculation. A key factor in the performance of DAF units is the size of the flocs. Larger flocs have a lower specific area (the surface area per unit of mass) than smaller flocs. If the floc is too large there is insufficient area for air bubbles to attach, and the flocs may not be removed. Thus as well as careful control of the saturated water flow, which affects the number and size of the bubbles, it is important to control flocculation to optimize floc size. Flocculant aids are therefore not normally used. The development of DAF is one of the factors that has led to a much greater awareness of the importance of coagulation and flocculation in water treatment in the UK. It is normal to have two-stage flocculation with variable speed flocculators prior to DAF.

DAF is a relatively energy-intensive process, because of the need to dissolve air into the supersaturated water. Typical energy requirements are 40–80 W h/m^3; a 10 Mld DAF plant would use 17–35 kW continuous power.

DAF is a high-rate clarification process with a short detention time in the flotation zone. Thus unless closely controlled the process is susceptible to upsets caused by changing raw-water quality or changes in chemical dosing. Nevertheless it is an increasingly common process. There are several reasons for this:

- the process is generally better at treating water containing algae and light flocs that settle poorly—indeed it requires smaller flocs than would be appropriate for clarifiers;
- the process starts up quickly and can be brought into use within 1 h of starting-up. Thus it is practical to use the process only when required;
- the sludge produced has a relatively low water content;
- space requirements are less.

Other processes. There are other clarification processes which may occasionally be encountered in water treatment. Examples include the Sirofloc process, which uses adsorption onto fine magnetite, other processes using adsorption onto other materials, and even coarse membrane filtration processes.

The patented Sirofloc process uses fine suspensions of magnetite (Fe_2O_3) to adsorb positively charged colloidal material. At low pH value the magnetite particles are positively charged, and when the magnetite is mixed with water containing negatively charged colloidal particles the particles are adsorbed onto the magnetite. The water, with the magnetite and the adsorbed material attached, is then passed through a magnetic field which causes the magnetite to aggregate into larger particles which easily settle. The magnetite is then recovered from the sludge and regenerated for re-use. The system claims to be very effective at colour removal from soft upland waters; however it has been little used in the UK.

8: Types of clarifiers

INTRODUCTION

Following coagulation and flocculation, water containing floc normally passes to a settlement/clarification phase. In the past the only process used was some form of settlement basin in which the flocs settled out. Nowadays it is common to use either DAF or some form of sludge-blanket clarifier; neither of these can be described as settlement basins; both are better described as clarifiers. This chapter considers both traditional settlement basins and the more modern clarification processes. It considers some of the practical aspects of clarifier selection and design and some of the various types of clarifier in more detail. There are two basic types of settling basins: horizontal flow and vertical flow; and two other types of clarifiers: vertical-flow solids contact tanks; and DAF tanks. First horizontal- and vertical-flow tanks are considered, followed by vertical-flow solids contact tanks; and DAF.

Horizontal- and vertical-flow tanks predominantly remove previously flocculated particles; there is some additional flocculation but this is not a major factor. Horizontal-flow tanks can be either rectangular or circular. Vertical-flow tanks are normally square hopper-bottomed tanks.

Vertical-flow solids contact tanks achieve much higher efficiencies partly by passing flow through a layer of sludge/floc. Flocculation and entrapment of fine particles as the flow passes through the sludge/floc layer means that these particles, which would otherwise rise and pass onwards, are retained within the floc layer. They have been developed from hopper-bottomed vertical-flow tanks, which are very deep, into flat-bottomed upward-flow clarifiers that have a relatively shallow depth of construction.

DAF tanks operate in a similar manner to horizontal-flow tanks, save that the particles rise rather than fall.

Pure theory is of limited use in designing a settling basin. For one thing it is difficult to predict the worst conditions under which the basin will have to operate. Although laboratory tests on a series of samples will give an indication of the most suitable types of basin and the required coagulant doses, and of optimum floc formation and settling velocity, generous factors of safety have to be allowed before the results can safely be applied in practice. Also although there may be extensive data on the performance of an existing plant which is being extended or modified, it is unusual to have anything other than relatively limited data for the design of new works: in some cases a new plant may be under construction at the same time as the reservoir which will supply the water to be treated. Where there is limited data for the design of a new plant, it may well be possible to use data from other plants in the region which treat water of similar quality.

CHOICE OF CLARIFIER

In the discussion of clarifiers that follows, the singular when used refers only to the type. No works of any size has less than two clarifiers and most large works have several, the number and size of the units depending in no small measure on the ease with which the chosen design can be scaled up.

The object is to choose a clarifier which will efficiently remove the floc, by settlement, or flotation, and from which sludge can be removed in an appropriate way. Some clarifiers are highly efficient but difficult to clean. Some are ideally suited to small works but do not benefit from scale effect and decline in merit on bigger works. Some work well when handled by experts but are unsuitable for unskilled operators. Some clarifiers will cope with higher silt peaks than others. Some are easy to build and some rather complicated. Some take up more space than others.

Although there is a wide choice, there is often one type of clarifier that suits a particular job. Some of the more common types are shown in Table 8.1. This table should be used selectively, as some of the columns may not apply in any given case. In general, the columns on the left will interest engineers in more sophisticated countries and those on the right will be significant in developing areas of the world.

This table has been simplified compared to the previous edition of this book to better reflect developments over the past few years. In the UK virtually all new water-treatment clarifiers use either

Table 8.1. Factors in selection of settling basins

	Standard of performance in normal conditions	Advantageous use of land area	Effectiveness with algae	Effectiveness on small works	Effectiveness on large works	Ability to withstand sudden changes in water quality	Ability to handle wide range of raw-water quality	Extent of use	Performance with unskilled operators	Low maintenance requirements
Conventional horizontal flow	+++	++	++	++	++++	+++	+++	++	++++	+++
Two level horizontal flow	+++	+++	++	+	+++	+++	+++	++	++	++
Radial horizontal flow	+++	++	++	+	++++	++	++	+	++	+++
Plate or tube settlers	++++	++++	+	+++	++++	+	++++	++++		++++
Hopper-bottomed upward flow	++++	++	+++	+++	++++	+++	++++	++++	++	++++
Flat-bottomed upward flow	++++	+++	+++	+++	++++	+++	+++	++++	+++	+++
DAF	+++	++++	++++	+++	+++	+	++++	+++	+++	++

Note:
Scale of + to ++++, low to high benefit.

flat-bottomed clarifiers with a sludge blanket, or DAF, preceded by a separate coagulation/flocculation stage. Elsewhere other clarifiers are used, and there is of course a wide range of different types of clarifiers in existing plants.

POTENTIAL PROBLEMS

There are a number of potential problems with clarifiers that designers should be aware of. These are discussed with respect to horizontal-flow tanks but many also apply to other forms of clarifier. Problems include:

- ensuring equal hydraulic loading of tanks;
- coincidence of peak output with peak turbidity;
- low temperature;
- excessive suspended solids;
- proper pre-treatment;
- liability to streaming;
- persistent wind;
- overturn of water in the basin.

These are discussed below. Times and figures mentioned normally refer to coagulated, well-flocculated water in which coagulant aids have not been used.

Ensuring equal-flow distribution between tanks

It is clearly foolish to design settling basins to operate at a certain loading and then to provide a design which leads to tanks operating at a range of loadings, with some of the tanks operating at higher loads than others. This is self-evident, but nevertheless it is common to find basins operating at different loadings due to poor hydraulic design. The best simple approach to splitting flows equally between tanks is to provide a proper flow-splitting chamber, or equivalent, using weirs with free discharge to divide flows to the tanks. Such an arrangement is simple and allows automatic redistribution of flows between remaining tanks if one tank is taken out of service. The drawback of such an arrangement is that it requires that a relatively high hydraulic head be available. This has to be provided for in the initial design of a treatment works; it is often the case that engineers are faced with an existing plant with poor flow distribution and insufficient hydraulic head available to introduce a flow-splitting chamber.

Where head is limited designers may provide a symmetrical layout that controls flow splitting by similar inlet losses in the feed arrangements using the treated-water outlet weir as a control. This is often unsatisfactory and a difference in flow of as much as 40% has been observed for six parallel tanks.[1] The problems with this method can be minimized by ensuring that losses in the inlet channels or pipes are low compared to losses in tank itself.

It is practical nowadays to accurately measure and control flows using magnetic flowmeters and actuated control valves. However, while this sounds attractive in practice it is complex and would not normally be considered except for very large works.

Coincidence of peak output with peak turbidity

In some countries, notably those with cold winters and hot summers, the peak demand can rise to very high seasonal peaks, often 50% or more above average. If this seasonal peak coincides with poor-raw-water quality and high suspended solids in a river it imposes very arduous conditions because the treatment plant is severely taxed at a time when maximum output has to be maintained. The designer has to pay particular attention to the sizing of clarifiers when such conditions occur. Whereas normally it may be possible to accept a degree of overloading to cope with short-term peak demands, this is clearly less acceptable if peak demands occur at the time of poorest water quality.

Low temperature

Settlement occurs in accordance with Stokes' law (Chapter 7): the downward velocity of the settling particles is inversely proportional to the viscosity of the water, which in turn is inversely proportional to the temperature. Thus, suspended particles sink more slowly in cold water. If poor raw-water quality occurs in the winter or spring, when water temperature is low the designer needs to allow for this. Where settling tests are carried out on samples of raw water, allowance must be made for raw-water temperature compared to the temperature of the water during the settlement test.

The temperature of surface water commonly falls to 0°C in the winter in many areas of Europe and North America. Maximum water temperature in the summer in the tropics can commonly be over 30°C. The dynamic viscosity of water at 0°C is 1.79×10^{-3} Pa s and at 33°C is 0.76×10^{-3} Pa s. Thus, there can be a two-fold improvement in clarifier performance in the summer compared to the winter. As water demands are temperature dependent, the variation in water temperature means that clarifiers normally perform better in the summer, a useful factor in coping with peak demands.

Freezing can also be a problem in cold climates and it is common to cover clarifiers where this can occur.

Streaming

The phenomenon known as streaming describes a condition in which some or all the incoming water does not mingle with the main bulk of water in the basin but passes rapidly through from inlet to outlet in a fairly well-defined stream. It occurs to some extent in all horizontal-flow basins but is particularly bad on radial-flow basins. Kamawamura[1] quotes an actual detention time of 30–40% of the theoretical in a rectangular sedimentation tank and tests have shown that in the worst instances some of the water entering a basin of 4 h capacity actually passes over the outlet weir within a few minutes. When this happens the whole theory of settlement is clearly inapplicable. This short-circuiting is due to large-scale eddy currents established either by some form of jetting, by the mixing of waters of different densities, or by wind. Generally at low loadings dead zones are larger and there is a greater degree of short-circuiting. Jetting is normally due to poor design of inlet arrangements; it is commonly associated with rectangular tanks and can be eliminated or minimized by proper design of the tank inlet and by using an appropriately proportioned tank. The design of horizontal tanks to minimize streaming is considered later.

Temperature differences of as little as 0.2°C and a turbidity in the raw water greater than 50 NTU can lead to the establishment of streaming.[1] Normally a sedimentation tank is warmed by the sun and the most common form of streaming is for incoming water, which is denser than the water in the tank as it is colder and more turbid, to drop to the bottom of a tank before rising at the outlet. Judiciously placed baffle walls will minimize large-scale streaming, but provision of a baffle wall often causes problems, including precluding the installation of mechanical scrapers. A basin with two compartments in series is a more effective answer. Some benefit also results from constructing basins that are long in relation to their width.

Where basins are exposed or where there is a steady breeze there can be problems arising from the wind setting up surface currents and eddies. This problem is common in coastal areas where there are sea breezes for much of the day. Settling basins are normally the highest point in a works, as the flow is often pumped up and then allowed to gravitate through the plant. The remedy is to screen or cover the basins.

Streaming is often very marked in radial-flow basins because the ratio of length of flow to width (i.e. $r/2\pi r$) is less than one. Vertical-flow basins suffer from streaming only if badly constructed with the outlet weir not level, causing unequal hydraulic loadings along the weir.

Excessive suspended solids

Aside from causing streaming, excessive suspended solids may give operational problems including excessive sludge production. Excessive suspended solids is not normally a problem in the UK but may be encountered in catchments where vegetation cover has been reduced and soils are easily eroded. The effectiveness of a basin declines if the incoming water contains excessive suspended solids. The maximum suspended solids that an upward-flow basin can normally take in its stride is about 900 mg/l. Rectangular horizontal-flow basins can treat water with higher suspended solids loads than vertical-flow basins. It would be unwise to expose the latter to suspended solids frequently in excess of 1000 mg/l, whereas the former normally cope reasonably well, especially if the hydraulic loading can be decreased by reducing throughput. Circular-flow tanks with rotating scrapers and high-capacity sludge lines can handle very high levels of suspended solids, of up to the order of 20 000 mg/l. However at such high solid-loading tanks would operate as pre-treatment, with high levels of solids in the settled water.

For simple tanks, a commonly used surface overflow rate of 18 m^3/day/m^2 together with a basin depth of 3–3.5 m gives a nominal retention time in a horizontal-flow basin of about 4 h. As a preliminary guide this might be varied up or down roughly in proportion to the ratio between the square root of the maximum suspended solids concentration to the square root of 900. For the typical fairly clear

UK river (of about 500 mg/l maximum suspended solids) this would give $(500/900)^{1/2} \times 4\,h$ and so permit the use of a basin of 3 h nominal retention capacity, whereas a fairly turbid tropical water, with 2000 mg/l suspended solids, would require about $(2000/900)^{1/2} \times 4\,h$ or about 6 h capacity. This rule of thumb has no theoretical basis but has evolved from many practical examples; it is applicable to situations where there is likely to be a poor level of operating ability.

Proper pre-treatment

Where settlement basins are being operated at high loading rates, near to their maximum capacity, it is particularly important that the water has been properly coagulated and flocculated prior to entering the settling basin. This applies particularly to where highly loaded vertical-flow solids contact tanks are used. These tanks depend in part on interaction between flocs in the incoming water and the floc blanket and have short retention times. It is important to maintain uniform water quality and a steady pH value for the process to operate consistently and a higher degree of coagulant control is appropriate

HORIZONTAL-FLOW SETTLEMENT BASINS

Introduction

At one time the rectangular horizontal-flow basin was the most widely used form of clarifier in the world. However now they are never used in the UK for new plants, and their use is unusual in new large plants in Europe. Rectangular horizontal-flow tanks are still used in North America because their advantages under typical American conditions outweigh their disadvantages. It is an extremely reliable basin and is very popular with plant operators because it rarely gives trouble and can always be relied on to outperform most other types when treating raw water with a high silt load and good performance is at a premium. One of the reasons for their continuing use in the northern USA is the ease with which they can be covered. Their proportions lend themselves to covering with reinforced concrete roof slabs to prevent freezing. In the US it is normal to have sludge scrapers, as access for cleaning is more difficult for a covered tank. In less extreme climates they are normally uncovered.

Clarifiers of this form are very easy to build and operate. The main advantages of horizontal-flow tanks are they are less susceptible to shock loads or process upsets and they are easy to operate and maintain. If they do not perform well they can often be upgraded by adding plate settler modules. They are not 'temperamental' and will put up with a lot of inexpert handling. Their considerable size makes it less likely that sudden fluctuations in raw-water quality will affect clarified-water quality. They 'scale up' very favourably and are at their most economic on big works. They are also able to treat waters with exceptionally high silt loads: they settle silt quite well, have room to store it and are not too difficult to clean. Their cost per unit of volume is low, and although they are bulky in appearance they are normally very cheap in overall cost. They are, therefore, very good performers on big works on silty rivers and can be operated by relatively untrained staff.

TYPES

Rectangular basins

In its traditional form a horizontal-flow basin resembles a large oblong box, filled almost to the top with water. The bottom is flat or has only a slight slope and the water is normally 3–4 m deep. Water enters at one end in the lower half of the tank and leaves at the other end over a surface weir. There may be baffles within the main box structure to inhibit short-circuiting. The basins typically have a retention time of at least 3 h, with longer retention times in colder climates or where there is no sludge removal equipment installed.

The rectangular basin is simple but has disadvantages: sludge collection mechanisms are not as simple as for circular tanks; and it requires more land than more sophisticated processes. Consequently efforts have been made to retain the merits of horizontal-flow while overcoming these problems. Radial-flow tanks (circular in plan), multi-storey tanks and tanks using shallow-depth tube or plate settlers are all variations on the theme of horizontal-flow tanks.

Radial-flow basins

There is no fundamental difference in hydraulic design between the rectangular cross-flow tanks and circular-shaped radial-flow tanks.

In a radial-flow basin the raw water enters through a central inlet and flows radially outwards towards a continuous peripheral outlet weir. The same values for surface overflow rates are applied and the tanks perform similarly to rectangular basins of the same loading, although as noted below short-circuiting is a common problem. Obviously, radial-flow velocities cannot be uniform because the cross-section increases with the radius, but this is not necessarily in itself a weakness, as maximum cross-section and therefore minimum velocity occurs where it is most needed, which is after the more rapidly settling particles have deposited.

Small circular tanks tend to be cheaper to build in concrete than square tanks of the same capacity, but they cover more ground per unit of area because of the wasted space caused by their shape. The outlet weir presents less of a problem because it can be placed right round the outer edge of the tank. Sludge collection uses rotating scrapers, which are relatively cheap and are very effective. However, streaming is a particular nuisance in circular basins because they are basically a poor hydraulic shape, length (i.e. radius) being less than breadth (i.e. perimeter). They are suitable in situations where there is a high silt load, which the rotating scrapers remove most efficiently. The Accelator type of circular upward-flow tank discussed below is superficially similar but has a far more complex design.

It is now unusual to find simple circular horizontal-flow settlement basins used for water treatment; where circular basins are used they tend to be complex proprietary designs incorporating flocculation and clarification enhanced by upflow solids contact in a floc layer.

Multi-storey tanks

It is quite possible to construct conventional horizontal-flow tanks in the form of a structure of two or more storeys. These often operate with the flow entering the lower level and flowing up to the top level, with the outlet above the inlet. Alternatively they may simply operate as two separate tanks, one constructed above the other. Two level basins are fairly common in some areas of the USA and in SE Asia. As area is such an important factor in settling, multi-storey tanks are often remarkably cheap and effective. Sludge removal is obviously more complex for a multi-storey tank, with a chain mounted flight scraper or some form of travelling suction device required. Where a tank operates in series scraping is sometimes confined to the first chamber, because most of the deposits form at the inlet end of the first storey.

Tube and plate settlers

Hazen and Camp arrived at the logical conclusion (cf. Chapter 7) that if area was the main consideration in settling tank design, the more extensive and shallower the tank the better. However, few (if any) very shallow tanks of great surface area have been built, but this principle has been adopted for the design of tube and plate settlers. Tube and plate settlers are increasingly common, both in new plants and installed to upgrade existing tanks. In principle there is no difference between tube and plate systems. Tube systems are normally hexagonal in cross-section, to permit them to be uniformly nested in blocks. They have the advantage of being more rigid than plates, which means that they deflect less under the weight of the settled sludge, and the rigidity is also useful for large units. They also prevent mass movement of water due to wind effects or temperature differences more effectively than plates. However, plates are easier to clean and are more common in water treatment. Plate systems are more flexible from the point of view of optimization as it is quite straightforward to adjust the design of the plate spacing to the optimum.

Specialist suppliers normally undertake the detailed design of such units, and there are many factors that affect the details of the units. They are normally fabricated from a suitable plastic or stainless steel. Plastic is lighter and is a low-friction material, which assists in sludge removal. The angle of the plates is typically between 50° and 60° to the horizontal. The smaller the angle the better the settling performance, but the poorer the sludge removal. Occasionally flatter angles may be encountered, but such units will either be co-current settlers, or will require periodic draining down for sludge removal.

Tube settlers are always set below the top water level in the tank, typically around 500 mm, with clarified water collected in a channel through submerged orifices or over a weir. Plate settlers may also be located below top water level, but some designs have the plates extending above water level, with the clarified water being taken sideways into collection channels. Because they extend near to the surface of water, plate and tube settlers tend to suffer from attached algal growth. This can be reduced by fabricating them from an opaque material to minimize light penetration into the unit, but can only be eliminated by covering the tanks.

A typical diameter for a tube is 75 mm. Plate spacing is typically 50–75 mm. Typical retention times in a purpose built unit are of the order of 15–20 min, but will normally be higher where an existing settlement tank is being upgraded. Overall surface-loading rates of up to 40 m/h are possible with tube or plate settlers. However, care must be taken in evaluating proposals to understand the basis on which the loading rate is calculated. For small installations, tanks will be significantly larger than might be anticipated because a high proportion of the plan area will be effectively unused due to the slope of the settling units (Fig. 8.1).

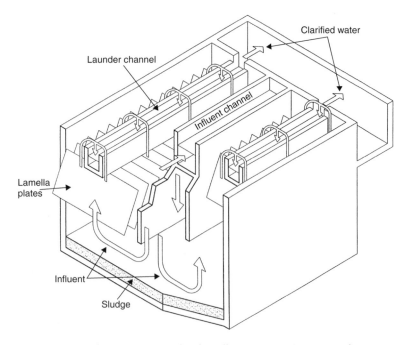

Fig. 8.1. Typical arrangement of a lamella separator (courtesy of Paterson Candy Ltd)

DESIGN OF RECTANGULAR HORIZONTAL-FLOW SETTLEMENT BASINS

The principles which govern the design of horizontal-flow basins have been covered in Chapter 7.

General

Disadvantages of rectangular horizontal-flow tanks include that they require a relatively large area and are prone to streaming. Thus the detail design of inlet and outlet arrangements is critical if a tank is to perform well. They do not benefit by the sludge accretion effect, and they need to be preceded by coagulation and flocculation if they are to work effectively. In cases where flocculation has been omitted, floc can often be seen forming part way along the basin. This is a sign of poor-process design and suggests that the performance of the clarifiers could be improved by adding flocculation.

As horizontal tanks do not depend on additional flocculation caused by particle collision, simple settlement theory is applied in their design. Of course as seen earlier flocculation does occur, even if it is not designed into the tanks. Tanks are designed to remove particles of the size necessary to provide a water of acceptable quality for the next stage of treatment, normally filtration. Acceptable filter loadings are discussed in Chapter 9, but a reasonable target water quality for clarified water around 5 NTU.

Loading rates

The first step is to determine the solids concentration in the raw water and the coagulant dose necessary to ensure formation of well-settling floc. This with the design flow rate will permit an estimate to be made of the quantity of solids to be removed by the tank and hence sludge volumes. Jar tests are essential to ascertain the optimum dosage of coagulant and the coagulation pH, the flocculation time, and the advisability of using coagulant aids. They will also indicate approximate surface-loading rates.

A well-formed floc typically settles at between 2 m/h at 0°C and 3.5 m/h at 20°C. Thus the simplest option is to design the tank on the basis of removal of such particles at the appropriate water temperature. Given that generally the highest water demands are in the hot

Table 8.2. Typical surface loadings for horizontal-flow tanks

Type	Q/A (m^3/day/m^2)		
	Very bad conditions	Normal conditions	Easy conditions
Without coagulant aids	9	18	24
With coagulant aids	18	27	36

conditions a value of 3 m/h is often used. This represents a surface loading of 3 m^3/m^2/h. In practice a safety factor of at least 2 is generally applied, meaning a surface loading of up to 1.5 m^3/m^2/h. This represents a reasonably conservative design loading for a good floc; rates of up to 2.5 m^3/m^2/h are quoted elsewhere but higher rates should be used with caution.

Table 8.2 shows the overflow velocities that are commonly used in designing conventional horizontal-flow tanks. Factors which favour settlement are coarse-grained sediment, high temperatures and low turbidity; factors which hinder it are colloids, cold water, high turbidity and the coincidence of peak turbidity with peak water demand. It is necessary to look at the worst conditions in each case and decide how bad or easy the situation may be at maximum works output, or alternatively how much water has to be produced when the raw-water quality is at its worst.

Alternatively tests could be made to determine the period required for the water to settle and a suitable design surface-loading rate. In practice full testing is rarely done but in the absence of plants treating similar water, some testing is clearly prudent if raw-water samples are available.

Dimensions of rectangular tanks

All other things being equal, rectangular basins perform better than square basins, because they are less prone to short-circuiting. Rectangular tanks with poor inlet arrangements are prone to streaming due to jetting of the incoming flow. This can be eliminated or minimized by proper design of the tank inlet and by using an appropriately proportioned tank. Kawamura suggests that tanks with a length to breadth ratio of less than two are generally ineffective,[2] and Cox[3] also suggests a length: breadth ratio of greater than two. Others[4] suggest a ratio of three, and where possible this is recommended as a minimum.

The depth of the settling zone is normally about 3 m. It is apparent that for a given depth D the detention time is directly related to the surface overflow rate (Q/A). A depth of 3 m is common for basins up to 60 m in length. Above that a depth/length ratio of 1 : 20 is commonly noted. In either case, if scrapers are not installed the depth should be increased to provide a volume in which sludge can accumulate before removal.

Retention times

Where the climate is temperate, silt loads are low, and there is mechanized sludge removal, horizontal-flow basins generally have a nominal retention time of around 3–4 h. A tank with a water depth of 3.5 m and a surface overflow rate of 28 m/day would have a nominal retention of 3 h, and a tank with a loading rate of 18 m/day would have a retention of 4.7 h. For river-derived supplies under typical UK conditions, nominal retention periods of the order of 3 h would be appropriate.

In many tropical countries, where the rivers tend to be more turbid, basins are commonly of at least 4 h retention capacity. In really difficult cases where there is very high turbidity, a high proportion of colloids or low temperatures, the basins may require a still greater nominal retention capacity but this would be most unusual. All the above retention times assume that coagulant aids are not used. For more difficult tropical sources, pre-sedimentation tanks may also be necessary.

If in doubt settlement basins should be designed conservatively as the extra capacity can normally be provided at low cost at the design stage and it can be difficult and expensive to correct for too small a basin. However, despite this, modern designs of horizontal basins do not normally have the very long retention times sometimes used for older designs, it being preferable to improve settlement by improving coagulation and flocculation and by using a coagulant aid.

Other factors

The water coming into the basin carries fragile floc which must not be broken up as it does not easily re-form. To ensure that the floc is undamaged the velocity in inlet pipes and channels should not exceed 0.6 m/s and should preferably be less under normal operating conditions. An appropriate range is 0.15–0.6 m/s, with the minimum velocity set to minimize deposition of flocs.

The design of the inlet to a horizontal-flow tank is critical to the performance of the tank and it is essential to ensure a uniform and steady-flow distribution across the tank if the tank is to operate to its full potential. Various arrangements can be seen in different text-books and there is no universally accepted arrangement. The best arrangement is to have an inlet channel running across the complete width of the basin. To ensure even-flow distribution into the tank, flow should enter, from the inlet channel, through a large number of ports distributed across the end wall of the tank. These should be 100–200 mm dia., with a maximum velocity of 0.3 m/s. The area of the inlet ports will typically be of the order of 5% of end wall area. Such an arrangement ensures a good distribution of flow at the inlet across the entire cross-sectional area of the tank. It normally removes any need for additional baffles, allowing easy installation of a sludge scraper. Where there is more than one basin there will need to be a separate flow distribution and isolation arrangement, to ensure equal-flow distribution between tanks and to allow one tank to be taken out of service.

Where the inlet arrangements are simpler, for example, using a weir inlet, a perforated baffle wall should stretch across the full width of the basin about 2 m from the inlet end. It should start just above surface and terminate about 1.5 m below. The velocity through any openings should not exceed 0.2 m/s. It should be possible for much of the water to enter the basin by passing down and under the wall. Kawamura[2] discusses the location and requirements for baffle walls in detail.

The average horizontal velocity in the basin should not exceed 0.02 m/s but generally the velocity will be well below this.

One of the most important features of any basin is the outlet weir, which is situated at the surface and has a length at least equal to the width of the basin. It has been found that short weirs, with high load-ings per unit length, induce currents leading to local high velocities, leading to poor settlement at the outlet end of the tank. In extreme cases at high flows there may be problems with scouring of deposited sludge from the floor of the basin. Making the weir a trough will combat this by increasing the weir length, with water entering over both sides and discharging sideways along the trough. In narrow basins two or more such troughs may be built. For a weir loading of 150 m^3/day/m, the maximum recommended, a single-lipped weir cannot discharge the water passing through a basin designed for a surface overflow rate of 18 m^3/day/m^2 if the basin exceeds 8 m in length, which in practice applies to virtually all horizontal basins. Multiple weirs of some sort, therefore, are normally desirable; they should be as near the outlet end of the tank as possible.

Sludge

As tanks have flat or gently sloping bottoms, removal of sludge cannot be done solely by hydraulic means but requires the installation of sludge scrapers to convey the sludge to hoppers. If sludge is voluminous, mechanical scrapers are necessary; in most countries they would be provided without second thought. However in older plants in parts of the world where labour and land are cheap, and where there are not high silt loads, horizontal-flow basins may be found without mechanical sludge removal installed. Such tanks require to be periodically drained down for sludge removal by hand. Where there is no mechanical sludge removal it is necessary to install additional treatment capacity to allow a tank to be taken out of service for cleaning while maintaining plant throughput; it is also appropriate to increase the retention time within the tanks to allow volume for sludge accumulation.

In long, narrow basins the sludge is scraped longitudinally into hoppers at the inlet end. Discharge is under hydraulic head and the sludge will normally have a solids content of 0.5–1.0% solids *w/v*, although it can be higher. To avoid settlement in the sludge pipes, velocities should exceed 1.4 m/s. Any new horizontal-flow tank should take into account the possibility of future upgrading by the addition of tube or plate settlers, and it would be prudent to provide sludge handling capacity additional to that initially necessary, to cater for this.

DESIGN OF PLATE AND TUBE SETTLERS

Practical design considerations are that the plate settler should be preceded by a coagulation/flocculation stage, most likely with addition of a coagulant aid. There should also be provision for removal of the coarser particles prior to their entering the settler unit. In practice this is ensured by providing a suitably designed upward-flow entry to the settlers; where any heavy particles will settle out. The plates or tubes should be placed so that the base of each bank is at least 1 m above the floor. Where the retention times are short it is essential that coagulation and flocculation are precisely matched to water quality. This reduces the probability of excessive carry-over of solids at times of changing water quality. If plate settlers are proposed for raw-water sources prone to rapid and significant quality changes, it is necessary to pay great attention to coagulation control and flocculation, and lower-loading rates and longer retention times are advisable.

For smaller units sludge can be collected in a hopper and removed hydrostatically. Larger units may use a series of hoppers or a scraper.

Tube and plate settlers are also attractive for upgrading the throughput of existing basins. In such applications it is usually only necessary to increase the performance of basins by a relatively small amount, although in theory it will be possible to achieve at least a four-fold increase in throughput, based on plate surface-loading rates. In order to utilize the full potential of tube or plate settlers in existing tanks it is often necessary to upgrade the feed and clarified-water pipes, and provide a large increase in sludge handling capacity.

Figure 8.1 shows a typical layout of a clarifier using a plate settler. An example of the design of a plate settler is given in Appendix 1.

TESTING OF WATER FOR THE DESIGN OF HORIZONTAL-FLOW CLARIFIERS

The first question is whether coagulation is required. This can be assessed by carrying out jar tests or by settling a sample of water in a tall glass cylinder. The cylinder is filled with the water to be tested without the addition of coagulants or stirring. In many quite heavily turbid waters the samples quickly show a clear dividing line between the upper clarified zone and the lower zone of settled silt. If this period is short and the line of demarcation is well-defined, flocculation is probably not necessary. If the period is lengthy and the zone of junction is blurred, colloids are probably present and flocculation is essential. In practice coagulation is normally required and jar-testing is necessary to define the conditions under which a well-formed and easily settled floc is obtained.

There is an argument for additional testing using a cylinder equal in depth to the proposed basin to take account of the interaction between settling particles. This would need to be a length of pipe of around 3 m long with sampling taps at different levels to permit samples to be withdrawn and analysed. However, in practice such testing is rarely done for several reasons:

- a large representative water sample is required and this may not be available—it is important to test the worst quality water likely to be treated;
- in nearly all cases flocculation would be carried out prior to settlement, so the water being tested would be coagulated and flocculated prior to testing; in practice it is difficult to do this in a laboratory for large quantities of water;
- the physical requirements for carrying out such testing are not simple, involving a settling column several metres tall; and
- such testing would almost always indicate a higher-loading rate than would be acceptable or used in practice.

Simpler, if time consuming, settling tests in a tall glass cylinder can be used to assess the settling velocity of the particles needed to be removed. After adding water to the cylinder and allowing it to settle for, say, 15 min a sample can be taken from the cylinder at a depth of, say, 500 mm. This can be tested for turbidity or for suspended solids. If the solids content or turbidity were satisfactory then the above figures would mean that removal of particles with a settling velocity of 2 m/h or more would be acceptable from a process standpoint. The test can then be repeated using a shorter or longer settling time to better define the minimum settling velocity of the particles that need to be removed. An appropriate factor of safety, normally three, should be applied to the results from laboratory testing in designing the full size tank.

VERTICAL-FLOW CLARIFIERS

Introduction

Initially upward-flow clarifiers were square-shaped basins, with an inverted pyramidal-shaped hopper. Flow entered the tank in the hopper and flowed upwards to outlet weirs around the perimeter of the tank. In theory any particle with a settling velocity greater than the upward-flow velocity would settle into the hopper and be removed. In practice such tanks were often more efficient than expected. The reason for this is additional flocculation caused by inter-particle collisions leading to improved solids removal. Thus it became usual to operate such tanks to maintain a blanket of fluidized solids in the tank to improve efficiency.

The principle of flocculation due to inter-particle collisions in a sludge blanket led to the development of more sophisticated upflow tanks of many patterns, which are often of proprietary design. Such tanks are often commonly referred to as sludge-blanket clarifiers but a more accurate description would be solids contact clarifiers, as their key feature is the flocculation that occurs due to particle collisions.

TYPES

Hopper-bottomed sludge-blanket basins

Figure 8.2 is a cross-section of a hopper-bottomed upflow basin. In these basins raw water (after addition of coagulant and flash-mixing) is admitted at the bottom of the inverted pyramidal base from where it passes upwards through a zone of fluidized settled floc (the 'blanket'). The passage of water through the blanket acts to flocculate and entrap the small particles in the feed water and greatly improves

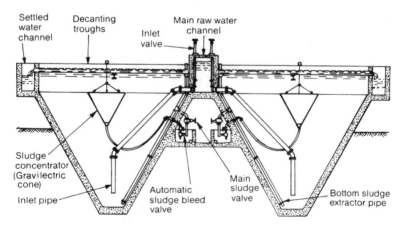

Fig. 8.2. Hopper-type settling tanks

Table 8.3. Typical surface loadings for hopper-bottomed vertical-flow clarifiers

Type	Q/A (m³/day/m²)	
	Low-alkalinity upland waters	High-alkalinity lowland waters
Without coagulant aids	20–40	50–80
With coagulant aids	30–60	70–120

clarification. Above the sludge blanket is a layer of clear water in the upper square-shaped portion of the basin; this makes it possible to observe the top of the sludge blanket. Under typical UK conditions such tanks work well. They are normally designed on the basis of the rate of rise permissible in the upper zone where the plan area is at a maximum. They were originally designed for, and are particularly effective on, water-softening plants. For such applications the permissible maximum upflow rate may be as much as 5 m/h.

In vertical-flow basins it is usual to maintain an upward water velocity of half of the normal settling velocity of the flocs. As well-formed alum floc settles at about 3 m/h under normal conditions, a typical upward-flow tank (without coagulant aids) is often designed to run with upward-flow velocities of 1.5 m/h. Table 8.3 gives typical ranges of permissible loadings for different sorts of waters, with and without coagulant aids. Where coagulant aids are used lowland river waters can be treated successfully at upward velocities of up to 5 m/h, although the higher values quoted would not be used without confirmation that they would produce clarified water of acceptable quality.

The great merit of these basins is that, under conditions that suit them, they provide both flocculation and settlement, and they give a settled water of great clarity. Sludge can be removed easily without a scraper, using only the available hydraulic head.

The drawback of these tanks is that, due to the typically 60° hoppers, as the sidewall length increases they become very deep, imposing an effective upper limit on their size. Thus large plants need a large number of relatively low-capacity tanks. The hopper-shaped bottom, which penetrates deep into the ground, imposes a limit on the economic dimensions of the square top (a limit which arrives more quickly when the ground is rocky or waterlogged). A tank of

5 m by 5 m will have an overall construction depth of the order of 6.5 m, but this will increase to around 10 m for a 10 m by 10 m tank. In practice bigger installations can be built only by adding an increasing number of fairly small units of standard size, each not exceeding 10 m square. While a reassuringly small total area of basin can still appear on the design drawings, it consists in most cases of a great amount of awkwardly-shaped concrete, and the cost is commensurately high. It therefore follows that the hopper-bottomed upward-flow tanks were most suitable for smaller works, of less than 45 000 m³/day.

The basins are normally considered unsuitable for treating waters with the high silt loads found so often in tropical rivers, because they are small and removal of large quantities of heavy sludge poses practical difficulties. They can be troublesome to operate when suspended solids exceed 1000 mg/l and if water more turbid than this occurs frequently horizontal-flow basins for preliminary treatment should also be used.

Another cause for concern regarding the use of hopper-bottomed upward-flow basins is that in hot countries in certain circumstances they have a tendency to 'roll over' in the afternoon. This is associated with a rapid increase in temperature of the incoming water and may be made worse by solar heating of the sludge blanket. The result is that the incoming water rises rapidly and the sludge blanket is also lifted. It is generally impossible to prevent changes in incoming water temperature, but shading the tanks will assist in some instances. The use of concentrator cones for sludge removal is also reported to alleviate the problem by permitting better control of the sludge blanket.

In older or unmodified tanks sludge was drawn off from the lower level of the cone into a chamber beside each basin using valves operated either manually or by a timer. Nowadays it is normal to use either sludge-blanket detectors to control desludging or sludge concentrator cones. Figure 8.2 shows a more sophisticated type of sludge draw-off, using a concentrator cone.

The concentrator cone is a cone made of a suitable material, often PVC-impregnated nylon. This is suspended within the clarifier with its rim just below the level required for the top of the sludge blanket. Excess sludge flows into the cone. The weight of sludge in the cone is measured, and when it reaches a preset value the drain valve on the cone is operated and sludge is allowed to flow from the cone.

This system permits accurate control of the sludge blanket and ensures the maximum concentration of solids in the sludge.

Modified hopper-bottom designs

The earlier forms of hopper-bottomed basins, as described above, had inherent weaknesses. They were expensive to build and difficult to operate on silty rivers. However, they operated on a principle that was basically desirable—the upward passage of water through a zone of mature floc particles which captured and retained the fine particles in the incoming water. Subsequent developments have resulted in evolution of advanced designs using the principle of passing the incoming water upwards through a sludge blanket; these operate very effectively while being far easier to construct.

Arguably the evolution of modern designs took place, although not in strict sequence, in four stages:

- the introduction of the multi-hopper tank;
- development of the basic flat-bottomed tank upflow clarifier;
- development of complex flocculator/clarifiers; and
- improvements and enhancements to the basic flat-bottomed tank.

The multi-hopper tank is a single tank with linked hoppers: it is equivalent to a battery of hopper-bottomed tanks with the upper common parts of the walls omitted. The advantage is a saving in civil construction costs. The disadvantage is that it hydraulically it is a single tank, rather than a set of individual tanks. This means that it is most important to equally distribute flow between the various hoppers. This is more difficult than dividing flow between separate tanks.

Flat-bottomed sludge-blanket clarifiers

The next logical step was the development of flat-bottomed tanks. For such tanks to work two problems have to be overcome: firstly incoming flow has to be introduced uniformly across the bottom of the tank, to ensure that there is a uniform upflow through the sludge blanket, keeping the particles fluidized; and there has to be a mechanism for sludge removal while maintaining the sludge blanket.

Early flat-bottomed clarifiers were normally circular. This enabled rotating sludge scrapers/collectors to be installed easily, to collect sludge into the centre of the tank. Such tanks would typically have a conical bottom, with a shallow slope towards the centre, but were much less deep than hopper-bottomed tanks. Flow was distributed and collected by radial inlet and outlet channels, with a multiple feed to the bottom of the tank. An alternative design from the USA utilized two channels running round the perimeter of the tank. The outer channel contained the incoming flow, which entered the tank through a series of drop pipes. The inner channel collected clarified water which entered over a weir.

It was difficult for simple flat-bottomed clarifiers to work well, but modern tanks are very effective. Two examples of modern tanks are shown in Figs 8.3 and 8.4.

Figure 8.3 shows a PCL design of a rectangular flat-bottomed clarifier. This is a true flat-bottomed design. Two important aspects of the design are the system for distributing incoming flow to the tank, and the sludge concentrator cones for removing sludge and controlling the level of the sludge blanket. PCL use 'Tridents' to take flow from a series of inlet channels running across the tank and distribute it equally across the tank. Clearly the detailed hydraulic

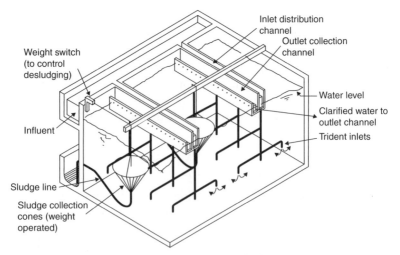

Fig. 8.3. PCL flat-bottomed clarifier (sludge blanket) (courtesy of Paterson Candy Ltd)

design of the flow-distribution system is critical for equal-flow distribution. The PCL 'Gravilectric' cones are used to collect, concentrate and remove sludge and to control the level of the sludge blanket.

Figure 8.4 shows an example of a Pulsator clarifier, made by Degremont. This is another type of upward-flow tank which depends on a sludge blanket for its effectiveness. It also has a flat bottom offering economic civil construction cost.

The Pulsator derives its name from the way that the feed water is admitted at varying rates of inflow. Incoming water goes to an airtight chamber from where it is fed to distribution pipework on the bottom of the tank. The inlet chamber is connected to a low-pressure vacuum pump. Air is withdrawn from the chamber at a rate higher than the maximum inflow of water. The water in the inlet chamber rises as air is withdrawn. The vacuum is then released, which causes a surge of water into the distribution pipework before the vacuum is applied again. Thus there is a period of low flow into the tank followed by a surge, followed again by a period of low

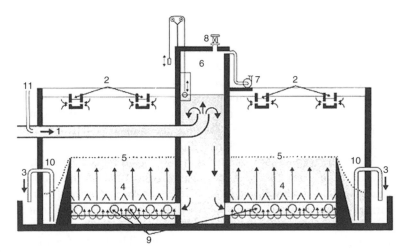

Fig. 8.4. Pulsator clarifier (courtesy of Ondeo Degremont). (1) Raw water inlet; (2) decanting troughs for clarified water; (3) sludge removal (controlled by valve); (4) stilling baffles; (5) upper level of sludge blanket; (6) vacuum chamber; (7) vacuum pump; (8) air release valve; (9) raw-water distribution system; (10) sludge thickening zones; (11) chemical addition

Table 8.4. Typical surface loadings for flat-bottomed sludge-blanket clarifiers.

Type of water	Q/A (m^3/day/m^2)	
	Low-alkalinity upland waters	High-alkalinity lowland waters
Without coagulant aids	30–60	70–120
With coagulant aids	50–100	100–200

flow. The cycle typically takes between 30 and 60 s. The sludge blanket expands during the period of maximum inflow and contracts when inflow diminishes. The movement induced keeps the sludge in suspension and encourages flocculation in the blanket. The tanks are generally rather deep, typically around 5 m. Sludge is decanted over a weir typically placed at about half tank height and can be withdrawn without significantly affecting the quantity of sludge in the blanket. To aid sludge withdrawal there are several draw-offs.

As in all upward-flow tanks, the water is collected by launders placed at regular intervals across the top to collect clarified water at a uniform rate across the tank. New tanks would normally be square or rectangular but the design can be retro-fitted in existing circular tanks.

The two design shown in Figs 8.3 and 8.4 are typical of the clarifiers being widely used across Europe at present. Such clarifiers also can be used with plate or tube clarifiers above the sludge blanket to allow even higher surface loadings. Typical loading rates are given in Table 8.4. The higher-loading rates quoted apply to clarifiers with lamella or tube settlers installed.

One of the potential problems with flat-bottomed clarifiers is that they are not easy to clean because the bottom of the tank is obstructed by pipes. While most of the sludge is removed during normal operation there is inevitably a build-up of heavier material in the bottom of the tank. It is therefore necessary to shut the basin down occasionally to permit manual cleaning. The frequency of this depends on the nature of the water being treated. Tanks treating soft low-turbidity upland water require cleaning less often than tanks clarifying lowland river water. Where there is likely to be a large volume of heavy sludge the use of other designs of clarifiers may be preferable, or pre-settlement basins could be used.

OTHER CLARIFIERS

There are many proprietary designs of circular clarifiers which combine flocculation and upflow solids contact clarification, and sometimes sludge recirculation. The various proprietary designs have a bewildering variety of names, the name generally intended to summarize the processes employed within the clarifier in a single word. Processes employed normally include some or all of: coagulation; flocculation; settlement; solids contact coagulation; sludge collection; and sludge concentration. The more complex are generally circular with an inner flocculation/coagulation zone and an outer clarification zone. Such clarifiers are relatively unusual in the UK for potable-water treatment but are more common in North America and Europe.

The simpler types consists of a circular tank inside which a smaller circular inner compartment acts as a flocculation chamber. The tank acts as a hybrid between a horizontal-flow and an upward-flow tank, dependent on the depth and diameter of the tank. The flocculated water passes downwards through the central zone and escapes radially outwards underneath the dividing wall (which is suspended from above and nowhere touches the floor). The outer, annular, compartment is sized to keep upward velocities at the levels acceptable for upward-flow tanks and the water is decanted from the top by radial collecting troughs. The floor is generally constructed with a slight slope towards a central drain. A centrally driven rotary scraper, with blades set at angles, pushes the sludge towards the drain, from which it can be withdrawn as required. The scraper sweeps the entire floor area, its horizontal arms being free to pass under the suspended dividing wall. The capacity of this sort of basin to handle heavy silt loads is one of its main features and they have been used to treat water from rivers prone to seasonally high silt loads

Accelator type solids contact clarifiers

The Accelator type of clarifier (Fig. 8.5) is a more complex clarifier also incorporating sludge recirculation.

It is circular in plan with a flat bottom and a diameter of up to over 30 m. It has a conical hood that divides the tank into two zones. In the inner zone, raw water and coagulants are mixed by a centrally mounted impeller before passing into the outer chamber, which acts as an upward-flow solids contact clarifier. The water enters the outer hopper-shaped zone and flows upwards to radial launders spaced at

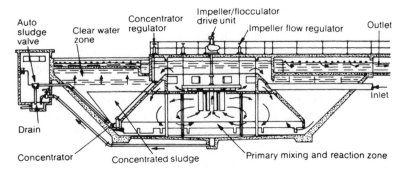

Fig. 8.5. Accentrifloc clarifier (by courtesy of Paterson Candy Ltd)

regular intervals to ensure uniform upward velocity. The surface area of the outer area is sized to provide the surface loading required.

The key detail of the Accelator design is that sludge passes from the bottom of the outer hopper into the central mixing zone. This recycling of sludge is induced by the movement of the liquid in the central section, which draws sludge in through the openings at the bottom of the hopper. The mixing of the sludge in the central area ensures high-efficiency flocculation. Within the outer hopper there are pockets in which sludge is collected prior to being taken from the reactor.

Clearly such basins are complex, with several variable process parameters within the single tank. Factors include rotation speed, solids concentration, as well as chemical dosing rates and pH value. They require more operator skill than simpler basins, and arguably are better suited to water from reservoirs, which is less prone to rapid changes in water quality than river water.

Water softening

Water softening is considered in Chapter 11.

DISSOLVED AIR FLOTATION (DAF)

DAF may either be provided as a separate treatment stage prior to filtration, or it may be installed above rapid gravity filters, combining

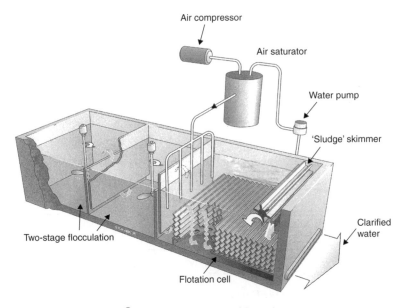

Air compressor

Air saturator

Water pump

'Sludge' skimmer

Clarified water

Two-stage flocculation

Flotation cell

Fig. 8.6. Purac Rapide® DAF unit (courtesy of Purac)

both clarification and filtration in a single tank. Figures 8.6 and 8.7 show examples of DAF units. Figure 8.6 is a schematic of a Purac DAF Rapide® DAF unit. The layout shown uses two-stage flocculation followed by a flotation cell. Sludge (foam) is removed by a rotating skimmer which sweeps it into a channel. The flotation cell has plates to ensure the water flows vertically downwards, with clarified water removed at low level.

Figure 8.7 is a schematic of the PCI CoCo DAFF plant, combining flotation and filtration in a single vessel. The upper part of the unit is a dissolved air flotation unit with the clarified water flowing directly onto a granular filter. There are two waste streams from the unit. The floated sludge is flushed off periodically at high level. Water is introduced from the flushing channel and flows across the top of the unit to the scum channel. Backwash water from the filters is taken off at lower level into the washout channel via the washout bay.

DAF plants operate at high surface-loading rates, within the range 6–15, typically $10 \, \text{m}^3/\text{m}^2/\text{h}$ although modern plants can operate in excess of $20 \, \text{m}^3/\text{m}^2/\text{h}$ where water quality is good and is stable.

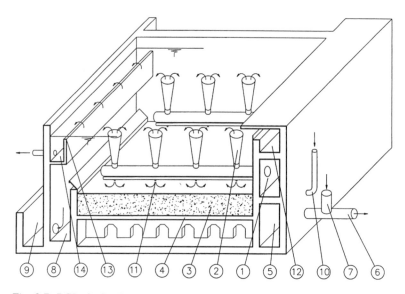

Fig. 8.7. PCL CoCo DAFF unit (courtesy of Paterson Candy Ltd). (1) Inlet duct; (2) inlet distribution cones; (3) filter media; (4) filter floor; (5) outlet and upwash duct; (6) outlet pipe; (7) upwash pipe; (8) washout bay; (9) washout channel; (10) air inlet; (11) air distribution nozzles; (12) flushing channel; (13) scum weir; (14) scum channel

Retention times are of the order of 20 min, less at high-loading rates, and tanks are typically 2–3.5 m deep.

APPROPRIATENESS OF CLARIFICATION PROCESSES

High design loadings are quoted for the many clarifiers of propriet-ary design. These can be achieved given good conditions, expert control and the use of polyelectrolyte. At the same time it should be appreciated that, even in plants operated by a highly skilled staff with automatic control and excellent laboratory facilities, often it is not possible to operate at the maximum loadings at all times, because of poor or changeable raw-water conditions. The higher-loading rates quoted earlier should only be considered for new plants where there is certainty that they are appropriate. Where there is doubt it is better to size for a lower loading, possibly providing for increasing the loading when the plant is expanded and the process has been proved.

It can be argued that the best solution at present for some developing countries lies with simple but well-designed works that can easily be operated by competent but not technically qualified staff designed on the basis of the lower rise rates given in Tables 8.2 and 8.3.

9: Filtration

INTRODUCTION

In a conventional water-treatment plant, the filtration stage was often considered the core of the process. The incoming flow to the filters was clearly non-potable, but after disinfection the filtered water was virtually potable water. As water-quality standards have developed this perception is no longer true in many cases, but filtration is still arguably the most important process in most treatment plants, and the process has been developed over the years to be more able to produce water able to meet higher water-quality standards. In the past filtration referred only to granular filtration, using sand or other granular material. Nowadays, where their use is appropriate, non-granular filters are becoming increasingly popular. This chapter mainly considers granular filtration in its various forms.

Basically the process of filtration consists of passing water through a granular bed, of sand or other suitable medium, at low speed. The media retains most solid matter while permitting the water to pass, and the filtrate from a filter performing well will be crystal clear with a turbidity of less than 0.2 NTU. Except for treatment of high-quality groundwaters, filtration almost always follows a clarification stage. Thus there is a trade-off between the performance of the clarification and the performance of the filters; the more effective the clarification the less the filters have to do and vice versa.

In order to achieve the required quality of filtered water, filtration has to remove particles far smaller than the sizes of the openings between the filtering media. Table 9.1 gives the relative sizes of sand grains and some of the particles present in water. A key requirement in water treatment is the removal of *Cryptosporidium* oocysts, which have a typical dimension of 5 μm, compared to a media particle size of 400–1500 μm (0.4–1.5 mm). Thus the processes involved in granular filtration are more complex than simple straining. The processes involved are discussed in more detail later.

Table 9.1. Relative size of sand grains and suspended matter

Material	Particle diameter (approx.) (μm)
Sand	800
Soil	1–100
Cryptosporidium oocysts	5
Bacteria	0.3–3
Viruses	0.005–0.01
Floc particles	100–2000

TYPES OF GRANULAR FILTER

There are three basic types of granular filter. It is useful to understand the differences between them before discussing them in more detail. The types are:

- slow sand filters;
- rapid gravity filters; and
- pressure filters.

Slow sand filters are the oldest form of filters. They are 'slow' because they operate at low-loading rates. They use fine sand as a medium and treat the water using two processes: physical straining and biological action. They have a layer of biological growth on top of the sand and in the uppermost part of the sand bed. This layer is vital to their effective operation, and until it develops slow sand filters provide only minimal treatment. These filters are cleaned at periods of between several weeks and several months, by scraping off the top layer of sand and biological growth.

Rapid gravity filters operate at far higher-loading rates. To do this they use coarser media with a higher permeability. They treat the water by physical treatment alone, although the media can be GAC, which also adsorbs chemicals dissolved in the water being filtered. Coagulation is normally required to ensure that smaller particles can be removed more effectively. Simple filters use a single medium, normally sand, but multimedia filters, which use two or more types of media, are common. Rapid gravity filters are cleaned by reversing the flow of water through the filters, to wash out the dirt, a process known as backwashing.

Pressure filters are a form of rapid gravity filters, the only difference being that they operate under pressure in large closed vessels. Traditionally they have been used on groundwater sources where water was pumped directly from a borehole directly through the filter into distribution, without any need for re-pumping. In such situations there was no clarification prior to filtration. They are sometimes used for surface water sources, normally on small plants.

The latter two types of filters depend solely on physical processes to remove solids, but slow sand filters also depend on biological action.

GRANULAR FILTRATION THEORY

The first important point about granular filtration is that simple straining is neither the most important method of solids removal nor is it desirable that it should be. If removal is predominantly by straining the result is that the filter rapidly becomes blinded, with an impermeable layer of matter on the surface of the sand. This results in the filter requiring frequent cleaning. This in practice means that there should be few particles in the water to be filtered larger than 20% of the size of the media. It therefore follows that it is important that such particles are removed during clarification, or where there is coagulation but no preceding clarification stage, that flocculation should not aim to produce large flocs, as these would very rapidly blind the filters.

Removal other than by straining is not straightforward. There are a number of mechanisms for transporting and attaching the particles to be removed to the media. Consider the idealized situation where the filter media is single size spherical particles. The first question is whether the flow is laminar or turbulent. The Reynolds number is given by:

$$Re = vD\rho/\mu$$

where D is the particle diameter and v is the velocity. However, where the flow is between particles with a porosity of f, the equation is modified to:

$$Re = vD\rho/(1 - f)\mu$$

Typically in water treatment D is of the order of 1 mm and v is of the order of 6 m/h. f is typically around 0.4, giving a Reynolds number of 2.8 at a temperature of 10°C. This is well inside the laminar range.

Transport mechanisms

Five different mechanisms have been identified[1] whereby particles in a flow of water through a filter may come into contact with a particle of filter media. Four of these are illustrated in Fig. 9.1. The five mechanisms are:

Interception.   Where a particle moving uniformly collides with a grain of filter media. The probability of this will be proportional to d_p/d_m, the ratio of the particle diamter to the media diameter. Thus the efficiency of interception can be increased by decreasing the size of the media, or increasing the size of the particle to be removed. For any given condition the larger particles will be removed more efficiently.

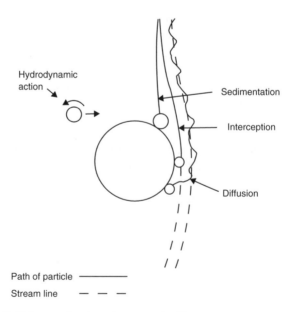

Fig. 9.1. Collision mechanisms in a granular filter

Sedimentation. This takes account of gravitational forces on the particles to be removed. For any particle the efficiency of removal will depend on the ratio v_s/v, where v_s is the settling velocity of the particle and v is the velocity of the fluid approaching the media. From equation (7.3), it can be seen that v_s is proportional to the density difference and the diameter of the particle squared. Thus dense particles will be removed more effectively, and larger particles will be removed very much more effectively.

Diffusion. This is Brownian motion, which affects significantly only very small particles. The number of collisions is proportional to $(T/d_p d_m v)^{0.67}$. As would be expected, the importance of diffusion increases at higher temperatures, and for smaller particles and media.

Hydrodynamic action. This arises from the velocity gradient in the vicinity of the grains of filter media. A small particle passing a grain of filter media will tend to be rotated by the velocity gradient. This will cause pressure differences across the particle, which will move it towards the filter media. This is believed not to be significant.

Inertia (impaction). The inertia of a particle heading for a collision with a grain of media will lead to a collision unless the hydrodynamic forces divert the particle to one side. At the low velocities and Reynolds numbers experienced in water filtration inertia is not considered a significant mechanism.

Attachment

Once a particle has made contact with the filtering media it needs to be retained. Provided there is no electrostatic repulsion this is effected by inter-particle attraction (Van der Waal's force), particularly for smaller particles. Where the particles are electrically charged there will be no attachment except to areas of opposite charge. It is important that charge neutralization in any preceding coagulation has been effective.

PRACTICAL ASPECTS

In practice for a given set of conditions the efficiency of particle removal varies with particle size. Figure 9.2 shows a typical variation

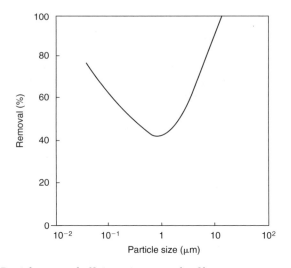

Fig. 9.2. Particle removal efficiency in a granular filter—typical

in particle removal efficiency with particle size. This shows that relatively large and relatively small particles are removed more efficiently than particles of around 1 μm. Very small particles are removed predominantly through diffusion and large particles are removed by straining. Particles around 1 μm are removed largely by interception and sedimentation, and these processes are less effective.

COAGULATION PRIOR TO FILTRATION

Filtration is used to remove the particles that are in the water. If the feed water is clarified water then the particles will normally be very fine, possibly with some flocs that have been carried over from the clarification stage. Sometimes, when the raw water has a low solid content the clarification stage may be omitted and the water pass directly to filtration. In either case it is common to dose a coagulant immediately prior to filtration. It is not sensible to dose a metal-based coagulant as this would itself impose a heavy solids loading on the filters, leading to rapid blinding and short filter runs. Thus the coagulant normally used is a polymer dosed at a low rate. Rapid mixing is required where the coagulant is dosed, but it is normally not necessary to flocculate prior to filtration as there is sufficient agitation between the dosing point and the filters for sufficient flocculation to take place, and large flocs are not desired. The coagulant used prior to filtration is commonly referred to as a filter aid. Filtration of water without prior clarification is often referred to as direct filtration. In the UK, where there is a risk of *Cryptosporidium* oocysts being present in the raw water, direct filtration is now often considered uncceptable as the sole main treatment process.

RAPID SAND FILTERS

Rapid gravity filters are the most common type of filter. They normally treat clarified water although sometimes they are used for treatment of high-quality raw water without prior clarification. Ideally they should be supplied with incoming water of less than 5 NTU of turbidity when they will produce excellent results. They may still perform reasonably well when the turbidity of the incoming water is 10–20 NTU but high turbidities will lead to much shorter run times and are not desirable. Operation at still higher turbidities may be possible but is unacceptable under anything but emergency conditions. A well-designed and operated rapid gravity filter should produce water with a turbidity of around 0.1 NTU. The feed water to rapid gravity filters is often dosed with a polymer coagulant to improve the removal of the fine particles carried over from clarification.

Cryptosporidium

In the late 1980s there was increasing concern over a number of incidents of cryptosporidiosis associated with the presence of *Cryptosporidium* oocysts in public water supplies. As a result the British Government established a group of experts under Sir John Badenoch to investigate the problem and recommend how best to address it. The Badenoch group produced two reports[2,3] which made a number of recommendations relating to the operation of water-filtration plants. Key recommendations relating to filtration were that:

- individual filters should be continuously monitored for filtered water turbidity and be backwashed in the event of the turbidity of the filtered water increasing;

- after bringing a filter back into service after backwashing there should be a slow start facility, and preferably an initial period of filtering to waste; and
- sudden changes in filtration rate should be avoided.

After an outbreak of cryptosporidiosis associated with groundwater in the late 1990s the 'group of experts' was re-established with Professor Ian Bouchier as chairman, to review development since the Badenoch reports. The Bouchier report[4] made several recommendations relating to water treatment, stressing the need to carefully monitor coagulation, clarification and filtration and to avoid by-passing treatment processes.

The recommendations arising from these three reports have greatly influenced water-treatment plant design and operation, particularly filtration processes, in the UK, and have meant that some filtration systems formerly used are no longer considered acceptable for new plants.

In 1999 the British government issued regulations requiring that any water-treatment plant considered at risk of having *Cryptosporidium* in treated water should either be continuously monitored for *Cryptosporidium* or should install a filtration system capable of removing particles down to 1 μm.[5]

Construction

Rapid gravity filters are normally constructed in an open concrete structure, with a building and pipe gallery at one end. The filter is basically a bed of sand, or other media, through which water is passed. This is supported on a bed of graded gravel. Beneath the gravel there is a system of underdrains that collects the filtered water and is also used to introduce air and water for cleaning the filter. The underdrain system has to collect filtered water uniformly across the filter during filtration and distribute air and backwash water uniformly during washing. These systems have to be carefully designed and constructed to ensure they do this efficiently.

Filter bottoms (collectors, nozzles, porous plates, etc.) and the general underdrain system are normally supplied by specialist manufacturers and differ widely in detail. The basic requirement is that they should consist of a large number of orifices, uniformly arranged to collect filtrate from, and deliver air and washwater to the underside of, the filter bed. At the same time they should not permit any filter media to pass. It is essential that the whole filter floor and nozzles be level. Although not essential, nozzle systems normally use a gravel layer to support the filter media. If the nozzles are designed with openings small enough to prevent media entering, it is necessary to have a very large number of nozzles to limit velocities and head losses through the nozzles; this is not normally practicable. The graded gravel supports the media while preventing media passing into the underdrain system. It also serves to equalize velocities across the layer, to ensure that there are no problems with high water velocities in the media adjacent to the nozzles. The nozzles typically have an opening of around 0.5 mm and are spaced across the floor of the filter to ensure a uniform-flow distribution across the filter. The detailed design of nozzles is quite a complex task and is normally done by specialist suppliers.

There are several different sorts of systems for the construction of rapid gravity filters. One common system uses a plenum underneath the filter floor for the collection of filtered water and the introduction of air and backwash water for cleaning. Figure 9.3 shows such a system whereas Fig. 9.4 shows a detail of a piped system. A plenum system provides an even distribution of air during cleaning and is less prone to operating difficulties than a piped system; it also allows access to the underneath of the filter. There are a variety of different systems used for rapid gravity filter underdrainage; earlier systems

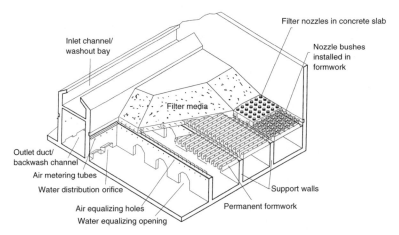

Fig. 9.3. Rapid gravity filter using a plenum floor (courtesy of Paterson Candy Ltd)

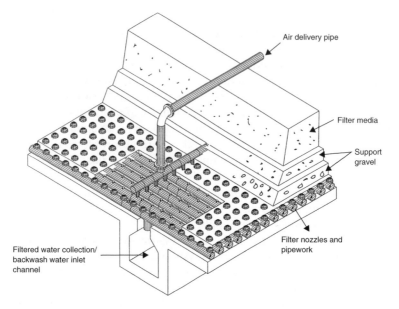

Air delivery pipe

Filter media

Support
gravel

Filter nozzles and
pipework

Filtered water collection/
backwash water inlet
channel

*Fig. 9.4. Detail of rapid gravity filter showing piped system of under drains
(courtesy of Paterson Candy Ltd)*

used filter tiles sitting on a solid concrete floor. Figure 9.5 shows a
modern equivalent of this system manufactured by USF Johnson.
This uses semi-cylindrical channels fixed to the floor of the filter.
The channels have a 'vee' section wire affixed, allowing an opening
of down to 0.15 mm. The channels are installed at a close spacing.
The combination of a small aperture and good coverage of the floor
means that the gravel layer is not required and sand can be placed
directly on the channels.

Normal practice now is to locate pipes in a pipe gallery at one end
of the filter, with a building above to house control and instrumentation equipment. The gallery will normally house some or all of the
following pipelines with associated valving:

- unfiltered water feed (more usually located at the other
  end of the filters);
- filtered water, normally including flow measurement;
- drain (connected to the filtered water outlets to allow
  filtration to waste);
- backwash water feed;

Fig. 9.5. Detail of Johnson Trition™ filter underdrain system (courtesy of USF Johnson)

- air feed;
- backwash water outlet (connected to a drain);
- sample lines (small diameter supplies for turbidity meters and particle counters).

The gallery will also contain flow measurement equipment and valves. As a consequence the pipework in the gallery is complex and crowded.

The size of individual filters is effectively limited by the backwashing process. Backwashing involves introducing a high upward flow of water through the media, normally in combination with air to break up the surface crust. The backwash flow is normally pumped. Large filters require very large pumps and pipes to handle the high flows and there are limits on the distance water can travel to a backwash outlet while maintaining effective cleaning. For these reasons filters rarely have an area greater than 100 m², with an upper limit of around 150 m². Depending on water temperature and media size, a filter of 150 m² can require a backwash flow of up to around 4000 m³/h, implying a backwash pipe diameter of around 800 mm. The proportions of the filter are affected by the need to ensure good flow distribution and to minimize the length of the backwash water-collection troughs.

Loading rates

The hydraulic loading rate on filters depends on the quality of the feed water, water temperature, and the media used. For many years normal loading rates for sand filters were around 6 m³/m²/h. Nowadays, typical rates are 6–8 m³/m²/h. Higher rates are possible with dual media filters (up to around 12 m³/m²/h), for coarse filters for manganese removal (up to around 15 m³/m²/h), or if the feed water is of a consistent and high quality. The loading rate selected has to take account of one filter in a plant being out of service and another filter backwashing. This situation would represent a normal worst case for loading. It is prudent not to design to the higher loadings quoted above without being sure that the filters will perform adequately under all circumstances, particularly given concerns over *Cryptosporidium* removal.

Filter hydraulics

A clean filter typically has a head loss of around 0.3 m under design conditions. Once the head loss reaches around 1.5–2 m the filter will be cleaned by backwashing. The depth of water over the filter when backwashing is required is governed by the downstream hydraulics and the need to avoid negative pressures developing in the media, see below. However, typically the downstream hydraulic control will be set at around the top of the media, and thus the depth of water over the media will be a maximum of around 2 m.

Development of subatmospheric pressure within a gravity filter

In the early part of a filter run when the media is clean the particles being removed are intercepted high in the sand bed, but as the top sand gets dirtier the particles penetrate more deeply and the loss of head increases. If the head loss in the sand at any point exceeds the

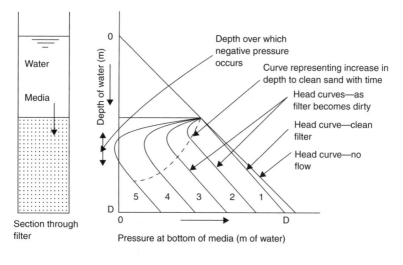

Lines 1 to 5 represent the changes in pressure through the filter as the media becomes blinded. Line 5 results in the development of negative pressures within the media.

Fig. 9.6. Development of negative pressure in rapid gravity filter

static head of water on the filter, subatmospheric pressure will be induced which may cause dissolved air to be given off from the water and result in air binding of the filter. In poorly designed or operated filters, this tends to happen at around the time that the filter needs washing. The reason for this can be seen from Fig. 9.6. The easiest way to ensure that this cannot happen is to have a level control on the filtered water outlet such that negative pressures cannot develop within the bed. The provision of an outlet weir with a cill level at the level of the top of the filter media achieves this.

Backwashing

Filters can often run for several days without washing. In modern plants the washing operation is automated and is triggered either: at predetermined times; by loss of head; or by excessive filtered water turbidity. The washing of a filter takes about 20 min. Normally filters are set to backwash at regular timed intervals, typically 24 h, but if there is a deterioration in feed water-quality backwashing may be triggered sooner by differential head loss, or by the turbidity of the filtered water. At the start of the backwashing cycle, the filter inlet is closed and the water level in the filter drops. After a short period a drain valve is opened to lower the level of water to the level of the backwash outlet weir. Once this level is reached, backwashing with air and/or water starts. Backwashing uses filtered water and is of key importance in maintaining filter performance. The particular backwashing regime adopted for a filter depends on the media used, filter layout, and the filter supplier's preferences. Backwashing may or may not fluidize the bed of media and may or may not use air as well as water.

Washing with water alone is not very effective in breaking up the crust which tends to form on the surface of the media. If washing is ineffective, the surface crust may crack and fragments can penetrate into the media to form mudballs. Once mudballs have become established in a filter they are difficult to clear. The use of air is particularly effective in preventing mudball formation and washing with water alone is now increasingly rare. In North America backwashing is sometimes combined with a surface wash system which directs water jets on to the filter surface during backwashing to break up the surface crust. Signs of concern in a filter are surface cracks, a tendency for the sand to shrink away from the concrete walls, and non-uniform surface turbulence during the washing process.

Initially rapid gravity filters were backwashed solely by pumping water up through the filter. For this to be effective the media has to be fluidized, typically involving a bed expansion of 15–20%, although in some circumstances expansion of up to 30% or more may be used. A 500 mm depth of sand could be expanded by up to 150 mm. The amount of water required to do this is dependent on the media size and density and water temperature. Lesser expansions are often quoted but these will not give fully effective cleaning using water alone. The backwash rate required for a given bed expansion is inversely proportional to the viscosity of water; thus it can be deduced from the viscosity data in Table 7.1 that at 20°C for a given bed expansion, a backwash rate approximately 50% higher will be required than for a water temperature of 4°C. Thus it is essential to be able to adjust backwash rates to take account of water temperature. Following a period of intensive washing, the backwash flow may be reduced for a rinsing period to carry dirt removed from the media into the backwash drainage system.

Modern filters use air and water for backwashing. The most common sequence is first to pump air through the bed to break up the surface layer, which has normally become blinded with filtered material. After a short time, typically 2 or 3 min, water is also pumped through the filter at a low rate to provide a limited bed expansion (typically less than 5%) and to carry dirt out of the media. This continues for around 5 min. The air scour will then be stopped and the water flow increased to rinse the media and carry the dirt into the backwash drainage system. Other sequences may be employed. Where a low-density media is used, air and water may be used separately. Multimedia filters and GAC filters should normally be fully fluidized during backwashing; this is to ensure that, if any mixing takes place, the media is restored to its initial position in the bed after backwashing.

Sometimes an additional flushing stage is employed at the end of the backwash cycle. This involves allowing clarified water into the filter above the media. The water is introduced via the filter inlet and it passes across the filter to the backwash outlet. Its purpose is to remove the dirty water above the media more quickly and effectively than can be achieved by the normal backwashing process. For this to be effective the backwash trough has to be located on the far side of the filter from the inlet.

When the filter has been washed, the filter is returned to service. As a result of the recommendations of the Badenoch reports[2,3] on *Cryptosporidium*, in the UK it is now considered essential to have slow start after bringing a filter back into operation, and where possible initially to filter to waste for a short period. This is because the filter takes a while to settle down, and the first water through the filter carries out some of the material disturbed but not removed during backwashing. In order to minimize the magnitude of the period of relatively high turbidity in the filtered water, it is essential to have efficient backwashing, and this has led to a tendency to lengthen backwash periods.

Flow rates during backwashing are very dependent on the bed expansion required, temperature, and the type and size of media. Typical airflow rates are $60–90\,m^3/h/m^2$. Water-flow rates can vary between 10 and $30\,m^3/h/m^2$. The amount of washwater used is typically about 1.5–2.5% of the total daily throughput.

It is during backwashing that the maximum practical dimensions of the filter assume importance, as a large flow of water has to be transported to, up and across, and then away from a filter. During backwashing the bed expands, lifting the surface. The weir over which the washwater escapes has to be higher than the bed level when expanded or there will be loss of sand.

Media

Traditionally rapid gravity filters have used sand as the main filtering media. This is then supported on layers of larger material. The size of the media is defined using three key parameters:

- *size range*—the maximum and minimum aperture sizes, the larger passing all the media and the smaller passing none of the media;
- *effective size*—the aperture size which retains 90% by weight of the media; and
- *uniformity coefficient*—the size of the aperture through which 60% of the media will pass divided by the size of the aperture through which 10% of the media will pass. This is an indication of the distribution of particle size in the media.

The other key parameter is the specific gravity, and there are also other parameters used to define the shape and porosity of media.

For single media filters the most commonly used media is silica sand with a specific gravity of 2.65 and a size of 0.5–1.0 mm. This normally has an effective size of around 0.5–0.6 mm and a uniformity coefficient of around 1.5. The sand should be sharp, hard, clean, and siliceous. The size of the media used will depend partly on filter loading rate and partly on the quality of the water being filtered. At higher-loading rates the effective size may be increased to 0.7–0.8 mm. The sand layer normally has a thickness of 0.5–0.75 m. A rule of thumb for the depth of media in rapid gravity filters[6] is that the depth of the media divided by the effective size should be equal to or greater than 1000. Thus if media with an effective size of 0.6 mm were used the bed itself should be 600 mm thick. The supporting gravel bed typically has a total depth of about 450 mm, making the total thickness of the filter media around 1 m. The supporting gravel normally consists of a layer of coarse sand or fine gravel on top of two or three layers of gravel; a typical size range is 2–50 mm.

A problem with single media beds is that during backwashing the bed is agitated and then allowed to settle. During backwashing the lighter particles tend to rise more than the larger ones, and when backwashing stops the larger sand particles settle faster than the smaller ones. As a result after backwashing the sand tends to become graded, with larger particles at the bottom and finer particles on top. Because of this characteristic, the filter does not follow the natural order of water-treatment procedure. In all other processes the bigger impurities are removed first. The single media filter, with its finest layer at the surface, tends to remove all impurities at one fell swoop. Clearly this is undesirable not only because the top layer of the media does all the work, but also, because the void spaces in the finer media are relatively small, there is limited capacity to store the entrapped impurities, and therefore the filter rapidly gets blinded. This is the opposite of what is required for long and effective filter runs.

There are several possible ways to overcome this. The problem would be reduced if the media were composed of identical grains, but in practice this is not possible. Another way is to have upward-flow filters, and there are examples found of such filters. The most common method of overcoming the problem is to have dual filter media; the combination of materials most commonly used is anthracite and sand. Anthracite has a specific gravity of approximately 1.6 and sand grains typically have a specific gravity of 2.65. This difference in specific gravities permits the use of anthracite particles larger than the sand used and means that provided the bed is fully fluidized during backwashing, the anthracite particles sink more slowly after washing and remain on top. The top layer of the filter is then composed of coarser media. There are no hard and fast rules for the thicknesses of sand and anthracite. Typical dual media filter would have 400–600 mm of anthracite of effective size 1.5 mm on top of 200–400 mm of sand with an effective size of 0.6 mm. The use of such dual media ensures a slower development of head loss with time as there is less blinding of the surface layer of media. Sometimes a third layer, of dense garnet, is included, but this is generally now accepted as having insufficient benefits to offset the additional complication and expense.

Apart from sand, it is quite common to find GAC used as a filter medium. This has the advantage of providing adsorption as well as filtration. The effectiveness of GAC as an adsorber is dependent on the contact time between the water and the GAC. In rapid gravity filters the contact time is relatively short, and this limits the effectiveness of GAC in filters. GAC adsorption is covered in Chapter 11.

While it is common practice to state the working rate of filters in terms of flow through a unit of surface area in a given time (for example, $m^3/m^2/h$), it should be remembered that filter performance depends essentially on adherence of the impurities to the surface of the sand grains themselves. The smaller the grain size of sand, the greater is the surface area per volume of media. It therefore follows that the depth of the sand bed and fineness of media also contribute to filter efficiency.

MODES OF OPERATION

It is normal to provide a minimum of six filters. This provides sufficient flexibility to cater for one filter being out of service and another backwashing. A minimum of four filters should be provided. Filters can be operated in several modes.

Declining rate filtration

Under this method of operation, all filters have the same inlet water level and the only control on filtration rate is the hydraulics of the filter media and pipework. Thus a filter that has just been backwashed will take a higher flow than the other filters. To prevent flow rates being too high there should be an adjustable valve on the outlet side of the filter to limit flow to a maximum rate, typically around 130% of the design average flow. When the filters are clean there will be a low head loss across the filter, and upstream water levels will be low. Over time the head loss across all the filters will increase and filtration rates fall slowly. Once the maximum possible head loss is achieved, filtration rates will decrease more rapidly. In order to allow for this it is necessary to provide either an overflow weir for excess water to be diverted away from the filters, or throttling of raw-water flow to limit flow to the capacity of the filters. The filters will be backwashed either at regular periods or when the minimum acceptable filtration rate is reached. There are drawbacks with this method of operation. When the filters are operating at design flow and with a feed water of constant quality, the system works well. However, for example, if the system is operating at 50% of design flow, a disproportionate proportion of the flow being treated will be taken by a newly cleaned filter. This is because although the flow will be restricted to 130%, say, of design flow, where the flow is 50% of design a newly washed filter could be taking 260% of the average flow. Thus when there are seasonal variations in flow, the hydraulic restriction on maximum flow should ideally be re-set. Also a deterioration in the quality of the feed water leads to an increase in the rate of head loss development, and this can cause problems if several filters then need to be washed to maintain throughput. Although used in the past, this method of operation is arguably no longer acceptable in the UK, not least because of concerns over the effects of high and uncontrolled filtration rates on removal efficiencies of *Cryptosporidium*. The use of declining rate filtration is discouraged in many US states.[6]

Constant rate filtration

Constant rate filtration is the normal mode of filter operation. This can be achieved either by control of the influent flow, or by control of the outlet flow. In either case the head loss through the filters increases as the filters become clogged.

Where influent control is used, flow is divided between the filters in service by means of inlet weirs. This ensures that all filters receive equal flows. Minimum water level in the filters is controlled by a weir on the outlet side. Over time the head loss across the filter increases and the water level in the filter increases. Filter backwashing is normally initiated on individual filters by either: the head loss across the filter reaching a preset level; filtered water turbidity reaching a preset value; or the filter having operated for a preset time.

Outlet control is used either to control the flows through filters, or to maintain a constant water level in the filters. Where it is used to maintain water level, flow is divided between filters by means of a weir inlet to each filter. The water level in each filter is then maintained at a more or less constant level by throttling a valve on the outlet. In the past this was done using a float-controlled valve, which led to some variation in water level dependent on the amount of throttling required. Now this is often done using a level detector to control a motorized valve.

However, the usual method for new filters is to provide flow control on the filter outlets to divide flow between operational filters. The flow into each operating filter is uncontrolled, with all filters having the same water level. Each filter outlet is equipped with a flowmeter and a flow control valve that is used to control flow through each filter. The bank of filters has a controller which calculates total flow and allocates it equally between operating filters. The controller also measures water level on the inlet side of the filters and maintains this at a constant level. Changes in flow rate, which would otherwise change the water level on the upstream side, are compensated for by increasing or decreasing flow through all the filters. Under this control philosophy the increase in head loss through the filter with time is balanced by opening the flow control valve more to compensate. This method of control requires accurate flow measurement for each filter and a PLC controller. It is the method most favoured at present. Having accurate flow measurement and a PLC controller provides a great deal of flexibility in filter operation. In particular it permits a controlled increase in filtration rate after a filter has been backwashed.

Performance

In practice a well-designed and operated rapid gravity filter will produce water of low turbidity for an extended period before there is increasing turbidity and the filter requires backwashing. After backwashing a filter will initially produce filtered water with a relatively high turbidity, of up to 0.5–1 NTU. This should drop rapidly to around 0.2 NTU or less within around 30 min and to around 0.1 NTU after a further hour or so. Figure 9.7 shows the shape of a typical plot of turbidity with time for the filtered water from an individual filter, as well as the development of head loss. Points to note are:

- When the filter is first returned to service after backwashing there is often high turbidity associated with the flushing out of material disturbed by backwashing that was not fully removed. The size of this peak partly depends on backwashing efficiency. Where possible it is now normal to initially divert the first flow of filtered water to waste to counter this.
- The filter will then produce low-turbidity water until it begins to rise as the filtering capacity of the media is approached. Once the turbidity reaches a preset value, or when the head loss across the filter reaches a preset value, the filter is backwashed.
- The intermittent loss of head peaks that can often be seen on a typical loss of head plot are associated with the backwashing

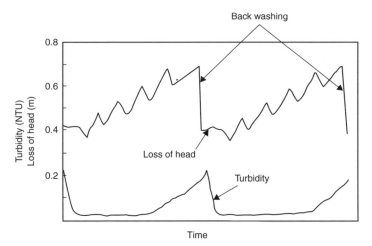

Fig. 9.7. Typical plots of filtered water turbidity and loss of head over time

of other filters. This results in a short term increased loading and may lead to increased turbidity in the filtered water, particularly when the filter is approaching backwashing.

The conventional rapid gravity filter consists of two basic systems: the filtering system and the backwash system. For modern plants there is also a sophisticated control system. In the UK, the costs and complexity of the current designs are accepted and are acceptable. However, there have been attempts to reduce the cost and complexity of rapid gravity filters, particularly for poorer parts of the world.

A common way of saving costs on the backwash system is to construct an elevated tank supplied either by the treated water main leaving the site if this is a pumped supply, or by a dedicated small pump if water flows to supply under gravity. This saves the cost and complexity of a large backwash pump. The pressure of the washwater required at the filter nozzles is about 0.6 bar. The top water level of the washwater tanks would normally require to be a minimum of about 9 m above the filter bed level. The tank would be sized to be able to wash two or more filters. A drawback is that large filters require very large flow rates, involving large diameter pipes from the elevated tank to the filters. Nevertheless for smaller plants providing backwash water from an elevated tank is economic and practical.

PRESSURE FILTERS

Pressure filters have many of the characteristics of the rapid gravity type but are enclosed in steel pressure vessels and are normally used where hydraulic conditions in the system make their adoption desirable. They can be installed, for instance, at any point in a pressure pipeline without unduly interfering with the hydraulic gradient, and their use often eliminates the need for double pumping. They are most commonly used where the water being treated is groundwater of a relatively good quality which does not require clarification. Often pressure filters on groundwater sources are used for removal of iron and manganese in addition to lowering turbidity. Where pressure filters are used on surface water sources they tend to treat good-quality soft upland waters from impounding reservoirs, which may also require iron or manganese removal. Such plants are common in the North of England. Pressure filters are also often used on smaller works or where simplicity of operation is advantageous.

There is no theoretical difference between the operation of a rapid gravity filter and a pressure filter. Rates of flow, criteria for washing and most other factors remain similar, although the maximum head loss through a pressure filter may be higher. Compared to rapid gravity filters, major differences are that it is not possible to see what is going on, flow splitting between filters is normally unregulated, and any chemical dosing and coagulation has to be done under pressure, in a baffled pressure vessel if needed. In a typical pressure filter installation there is a noticeable absence of pre-treatment and flow controllers and, as these have considerable value, it follows that pressure filters must be working at a disadvantage.

The filter shells are commonly of 2.4 m dia., but can be up to 3.5 m dia., and can be installed vertically or horizontally. Horizontal units (Fig. 9.8) are normally encountered in large installations where the smaller sand area in a vertical shell ($5 \, \text{m}^2$) would necessitate too many units. The length of a horizontal unit does not normally exceed about 15 m and thus the unit has $2.4 \times 15 \, \text{m}^2 = 36 \, \text{m}^2$ of media area.

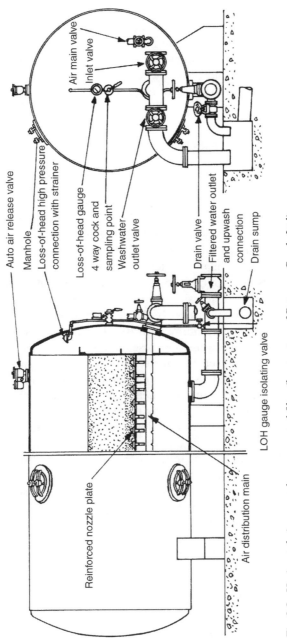

Fig. 9.8. Horizontal air-scoured pressure sand filter (by courtesy of Paterson Candy Ltd)

Theoretically the round shell of a horizontal filter constricts the disposition of the underdrains and detracts from the efficiency of the filter, but this is not significant in practice.

Pressure filters are suitable for the removal of iron up to around 10 mg/l, and for removal of low levels of manganese and are commonly used for this in small- or medium-sized plants. This is covered in Chapter 11.

SLOW SAND FILTERS

Slow sand filters are the original type of sand filters and have a successful history going back to the early years of the nineteenth century. Slow sand filters were first used in London in 1820 and there are still many big cities (including London and Bristol) where they still treat much of the water supplied. They work without coagulants and are often found on reservoir or lake-derived supplies. They can produce excellent quality water, but use is declining because the process takes up a lot of space and can be labour-intensive to operate. They can work for short periods or at low-loading rates with incoming water of up to 30 NTU turbidity but ideally require an average turbidity of not more than 10 NTU. The slow sand filtration process has been extensively researched recently and a notable improvement has been the inclusion of a layer of GAC within the media. Slow sand filters have considerable merit, and in countries where land and labour are cheap, and chemicals are relatively expensive; the possibility of using them should not be overlooked.

The principle of operation of slow sand filters is different from that of rapid filters, the main difference being that the filtering action mostly takes place at or near the surface of the sand. On the sand surface, a biologically active mat comprising: debris from the water being treated, algae, protozoa, bacteria, and other organisms, is established. This is known as the schmutzdecke (sounding more impressive in German than its English translation of 'layer of dirt'). It is within this mat that much of the treatment takes place, with suspended and dissolved matter including micro-organisms being removed by both physical and biological action. However, treatment extends 300 mm or so into the filter, in a heterotrophic zone, where organisms use the nutrients in the water as a source of energy. Thus slow sand filters are emphatically not mere filters and they do not provide acceptable treatment until the schmutzdecke has become established. A mature slow sand filter also has attached algae growing in the water above the sand and there is some treatment here, before water passes through the filter.

A slow sand filter is brought into service by filling it with water. This is done by filling upwards, introducing water beneath the bed in order to reduce the risk of air being trapped in the fine sand. It is then operated at a low rate for a period of between 3 and 10 days, with the filtrate being either discharged to waste or recycled to the head of the works. During this period the schmutzdecke forms and 'ripens'. Once this has happened the filter can be brought into normal service. In the summer when there is more light and the water is warmer the filter ripens more quickly.

The filter typically will then operate for a period of between 2 weeks and up to 3 months, the length of the run depending on the quality of the water being treated, the time of year, and the filter loading rate. In the summer runs are much shorter than in the winter due to there being more light and a warmer water temperature, both of which encourage algal growth. Covering a filter to reduce the intensity of the light reaching the filter dramatically increases run times. Cleaning is carried out by skimming a depth of around 20 mm of sand off the top of the filter. This can be done by taking a filter out of service, draining it down, and then using plant or hand labour to skim off the sand. Alternatively a mechanized skimming and suction system that does not require the filter to be drained can be used. After cleaning the filter is then restarted as described above.

Once the thickness of the sand layer has been reduced to 400–500 mm by repeated cleaning, the filter has to be re-sanded. This is done by removing around 150 mm of sand and then replacing sand up to the design maximum thickness. Sand removed from the filters can be washed and re-used if economic.

Slow sand filters are of simple construction (Fig. 9.9). A deep bed of sand rests on graded gravel, within which underdrains of open-jointed tiles are buried. The sand is carried to the full depth of the bed near the outer walls and no pipes should be laid within 0.6 m of the walls, to avoid water short-circuiting down the walls and through the gravel to the underdrains. The sand is normally finer than that in a rapid filter, and its quality and grading are less exacting, (Table 9.2), so it is more likely to be found locally. It is vital to have sufficient filters to ensure that when a filter is out of service the

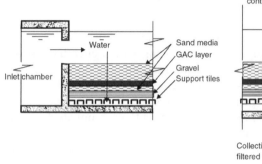
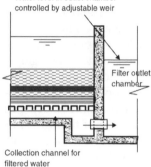

Fig. 9.9. Slow sand filter

Table 9.2. Typical slow sand filter construction

	Depth (m)	Grading (mm)
Water	1.2–1.8	—
Sand	0.8–1.2	0.2–0.4
Fine gravel	0.05	5–10
Medium gravel	0.05	10–25
Coarse gravel	0.15	10–80
Underdrains	—	—

increased loading on the remaining filters is acceptable. Except on very small works there should never be fewer than four filters, and six is a reasonable minimum number.

Each filter requires a method of measuring and controlling the flow through that filter. From a control point of view it is easiest to measure and control flow on the outlet side, with the inlets of the filters all at the same water level. If inlet valves were to be used for controlling flow through the filters there would be problems caused by the long response time; this is because both the water level above the filters and as a consequence the flow rate through the filter would change slowly in response to changes in incoming flow. Manual control is quite practicable because of the slow development of head loss. If flow measurement is not provided then the alternative is weir inlets to each filter, which under normal operating conditions would provide an equal loading to each filter. During ripening of a filter, throttling back the outlet valve would cause the water level in the filter to back up and drown the inlet weir. On the outlet side, there needs to be provision for diverting flow to waste during the ripening period and for draining down the filter. There also needs to be provision for filling the filters with water via the underdrains.

As cleaning the filter can mean taking a filter out of service for up to 2 weeks, if manual cleaning is used, it is important to monitor filter status carefully and plan cleaning so that throughput can be maintained, particularly in the summer when demand is normally highest and filter runs are shortest. Taking a filter out of service immediately increases the loading on the remaining filters and the danger is that this will then mean that another filter will then require cleaning. Because of the significant time lag between taking a filter out of service and it being back in operation there is a danger of filters cascading rapidly out of service, causing major production problems.

Traditionally slow sand filters have been loaded at a rate of around 2–5 m^3/m^2/day (0.08–0.2 m/h). However, if the raw water is of a high quality and provided they are managed properly they can operate at average rates of up to around 12 m^3/m^2/day. The slow sand filter works serving London have been upgraded by using rapid gravity filters as roughing filters for pre-treatment, and the slow sand filters now operate at this high rate at times of peak demand. To achieve such high loadings a turbidity of less than 5 NTU is required for the water being treated.

The head loss through a clean filter is small, as little as 50 mm. The maximum head loss before cleaning is required is normally taken as around 1 m. Adjusting the outlets to increase the head available over the sand provides the increased head required as the filters become clogged.

The merits of slow sand filters are that the quality of the filtrate is high, well below 1 NTU, and essentially all harmful bacteria are removed; no chemicals are required or even desirable; the filters can be built with local materials often using local sand; and operation is easy. The need for pre-treatment depends on the quality of the raw water and the loading rate applied to the filters. Where filters are to be operated at high-loading rates or will treat waters with high turbidity or suspended solids, pre-treatment is required. This is normally microstrainers or rapid gravity filtration to remove turbidity but where directly abstracted river water is used a settlement stage may be necessary to remove silt. Water with a turbidity of up to 30 NTU can be put directly on to the filters at rates up to 5 m/day. (Such values are higher than for most reservoir-derived waters, for which slow sand filters are specially suited.)

The disadvantages of slow sand filters is that they cover a large area and land costs may be high; they do not deal effectively with highly turbid water, nor will they remove colour; algal growths cause operational trouble; and unless sand skimming is mechanized, the filters are labour-intensive.

A significant improvement to slow sand filters has been to include a layer of GAC within the sand layer to adsorb organic chemicals. This would typically be around 150 mm thick, providing a contact time of 30 min at a loading rate of 7.2 m/day. This is normally sufficient for lowering pesticide and colour to acceptable values. The GAC has to be replaced at regular intervals and this would normally be done when the bed is re-sanded. A further possible improvement is to cover the filters. This greatly extends filter runs.

Slow sand filters have undergone a dramatic reassessment over recent years. They have long been recognized as very suitable for use on reservoir- or lake-derived supplies for small communities where technical supervision is lacking. However, they tended to be regarded as old-fashioned technology, requiring large areas of valuable land. Recently this impression has changed. By upgrading pre-treatment they can be operated at higher rates and by including GAC in the media water suppliers can utilize them for pesticide removal, avoiding the cost of constructing dedicated GAC adsorbers. In the Netherlands slow sand filtration is used as a polishing stage in plants designed to produce biologically stable water with a very low biologically assimilable organic carbon content.

OTHER FILTERS

Traditionally water-treatment plants have used the filters considered above. It is likely that rapid gravity filters will continue to be the filtration process used in most treatment works. Such plants are flexible and well understood and trusted. There are, however, a large number of other forms of filter that may be experienced. Not all of these are approved for potable-water treatment in the UK.

The 'Dynasand' filter is a sand filter employing a moving bed of sand. This is constructed as a steel cylinder. Clean sand is introduced at the top of the filter. Water to be filtered enters at the bottom of the filter and flows upwards. At the same time the bed of sand is slowly moving down, with dirty sand being removed from the base of the filter by an air lift pump. This sand is cleaned and returned to the top of the filter. These filters are suitable for pre-treatment of water but are not acceptable in the UK as the main filtration process. They can operate at rates up to around $15 \, m^3/m^2/h$.

A system that offered significantly lower capital costs used small cells and a travelling bridge to backwash individual cells. The filter was constructed from cells approximately 500 mm wide and typically 10–15 m long. The number of cells was set by the filtration capacity required. Backwashing used a travelling bridge on which was mounted the backwashing pump and a hood the same size as a single cell. To backwash the filter the hood was positioned over the first cell and the pump started. The pump drew water upwards through the cell and discharged into a backwash channel at one end of the filter. After several minutes of backwashing the pump was stopped, and the bridge moved on to position the hood over the next cell. The pump was then re-started and that cell washed. During backwashing the rest of the cells continued to filter as normal. The system has several problems: there is no control over filtration rates of individual cells; it is not possible to filter to waste or provide a slow start after backwashing; and there is concern that at the end of backwashing raw water from above the filter may pass directly through the filter as the bed settles. For these reasons, although the system is used successfully at a number of drinking water plants in the UK, it is generally no longer considered acceptable for new plants, because of concerns over the opportunity for *Cryptosporidium* oocysts to pass through the filter cells during and immediately after backwashing.

Other filters use mats or fibres to remove fine solids. One design uses a large number of fibres which are twisted around a central cylinder. Flow passes through the mat of fibres into the cylinder. The efficiency of such systems improves as material is deposited on the filter. When the head loss reaches a preset value filtering is stopped and the filters are untwisted, allowing dirt to be flushed from the loose fibres. A filter of this sort can remove particles down to below 5 μm, achieving better than a 99% removal of *Cryptosporidium* oocysts.

Diatomite filters are compact, high-efficiency filters which are suitable for armies in the field, and swimming pools, and for meeting short-term emergencies. They are small and portable and depend on the deposition of filter powders of diatomaceous earth (Kieselguhr) on porous filter 'candles' for their filtering action. They cannot deal with highly turbid water and because of extremely high head losses in the filter their running costs are high.

10: Membrane processes

INTRODUCTION

This chapter gives an overview of membrane processes. It will define the various membrane processes, introduce some of the key concepts involved and discuss the reasons for their increasing use.

The term membrane process applies to processes that use membranes to remove either very small particles or molecules and ions from water. Figure 10.1 shows the range of particle sizes associated with membrane treatment and the membrane treatment processes used for different sizes of particles.[1] It should be noted, however, that there are no hard and fast definitions of the particle sizes treated by the different processes and different definitions to those given below are common. It is also common to specify the approximate molecular weight of the molecules retained by the coarser membranes as this better reflects their performance.

Reverse osmosis is the process that removes essentially all particles and dissolved chemicals from water. However, small dissolved undissociated molecules and dissolved gases do pass through the membranes. The mechanism of separation is a mix of physical straining and diffusion. Although it is arguably misleading, reverse osmosis is often considered to remove particles below 1 nm (0.001 μm). Nanofiltration is reverse osmosis but refers to the use of leakier membranes that typically allow through a high proportion of monovalent ions and a small proportion of bivalent ions. With the wide range of membranes available reverse osmosis membranes can be optimized to remove particular sizes of dissolved molecules and ions and the use of the term nanofiltration implies only that the smallest and most mobile ions will pass through.

Ultrafiltration is largely a straining process, essentially a very fine sieve. It removes suspended matter, colloidal material and, where the pore sizes are at the lower end of the range, some large organic molecules. It is considered to remove particles in the range of 0.001 to around 0.1 μm.

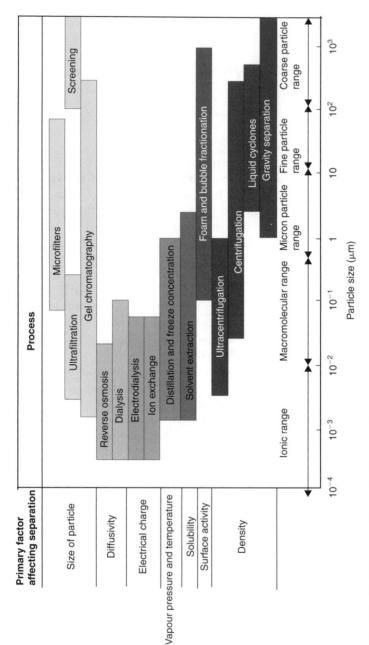

Fig. 10.1. Particle sizes with applicable separation processes

Microfiltration is a pure straining process; removing particles in the size range 0.1–100 μm without affecting the soluble materials in the water being treated.

Another membrane process is electrodialysis. The key difference of electrodialysis is that it is an electrical-based process that only removes ions and charged particles, although there may be some secondary straining.

REVERSE OSMOSIS

Perversely it is easier to define osmosis before attempting reverse osmosis. Osmosis is the term for the phenomenon whereby if two salt solutions of different concentration are separated by a semi-permeable membrane, water will migrate from the weaker solution through the membrane to the stronger solution, until the solutions are of the same concentration. The salt will be retained by the membrane and will not cross from one side to the other. Reverse osmosis involves applying a differential pressure to reverse this natural flow, forcing water to move from the more concentrated solution to the weaker.

The pressure required to do this has two components. The first is the pressure needed to prevent water moving to the more concentrated solution. This is the osmotic pressure and is a function of the difference in ionic strengths either side of the membrane. Assuming a temperature of 27°C and pure water on one side of the membrane this pressure varies linearly from approximately 0.4 bar for water with a sodium chloride concentration of 1000 mg/l to approximately 14.8 bar for water with a sodium chloride concentration of 35 000 mg/l (approximately equivalent to seawater). This pressure is the minimum required but to it needs to be added head losses through the membrane. The pressure due to head loss through a membrane depends on the membrane used and the flow rate per unit of area through the membrane. For a given membrane material it is inversely proportional to the thickness of the membrane. It is typically of the order of 15–60 bar, with more saline waters normally requiring higher pressures. Combining the two components it can be seen that the flow of water through a membrane is given by:

$$Q_w = K_w \times (A/t) \times (\Delta P - \Delta \pi). \tag{10.1}$$

Where:

Q_w = water-flow rate

K_w = membrane permeability coefficient for water

A = area of membrane
t = thickness of membrane
ΔP = pressure across membrane
$\Delta \pi$ = osmotic pressure.

Membranes used in reverse osmosis are not completely efficient semi-permeable membranes; they also allow some diffusion of salt. This depends on diffusion and the flow rate is given by:

$$Q_s = K_s \times (A/t) \times \Delta C_s. \tag{10.2}$$

Where:

Q_s = salt-flow rate
K_s = membrane permeability coefficient for water
A = area of membrane
t = thickness of membrane
ΔC_s = difference in salt concentration across the membrane.

The two important points that follow from equations (10.1) and (10.2) are that:

- increasing the differential pressure increases water flow across the membrane;
- the rate of salt movement across the membrane is independent of pressure.

Thus from the point of view of minimizing treated-water salt concentration it is better to operate at high differential pressures (which maximize water flow) and low differential salt concentrations (which minimize salt flow across the membrane). However, energy and pre-treatment costs also need to be considered, as will be discussed later.

Membranes

There are a variety of membrane materials and development of new improved materials is continuously happening. Initially membranes were based on cellulose acetate. This was developed into a range of cellulose-based polymers. A notable advantage of these membranes is that most are tolerant of chlorine at low levels, meaning that bio-fouling can be avoided; this is most significant as they can be biologically degraded. Other materials have subsequently been developed notably polyamide-based materials. There is now a wide range of polyamide materials and the properties of the material can be optimized for particular applications. Of particular note are thin film polyamide-based materials which being thinner, consequently with less head loss across the membrane, operate at significantly lower pressures. The polyamide-based materials are resistant to biological degradation, but are chemically attacked by chlorine.

The two most common forms of membranes used are spiral wound modules and hollow fibre modules. In both cases the material used will comprise the membrane itself and a more porous supporting material to provide the necessary strength to allow the module to be manufactured.

Spiral wound modules are of a Swiss-roll construction, consisting of three layers:

- a high-pressure layer which contains the feed water which is to be treated;
- the membrane layer;
- a low-pressure layer which contains the treated water, the permeate.

The high- and low-pressure layers contain spacers which permit flow to pass through them. Normally the permeate flows to a collection tube in the centre of the module. Feed water is introduced at the outside of the module and the concentrated feed water, the concentrate, is collected at one end of the module.

Hollow fibre modules have the membrane in the form of a hollow fibre. These have the membrane layer on the outside of a hollow fibre of supporting material. Flow is from the outside of the fibre into the hollow centre. The fibres are sealed into pressure bulkheads, normally forming a 'U' tube within the module. The feed water and the concentrate are on the outside of the fibres and the permeate is collected from where the fibres are sealed into the pressure bulkhead.

Pre-treatment

Given the very small pore sizes involved it is vital that water is adequately pre-treated prior to the reverse osmosis process. Furthermore, hardness is a particular problem as it leads to scaling. This occurs where the concentration of calcium carbonate or calcium sulfate in the concentrate increases to a value higher than the solubility limits of these chemicals.

Pre-treatment involves producing a low-turbidity water using coagulation, settlement, dual media filtration, and either cartridge filtration or ultrafiltration, to remove all particles greater than 5 μm or less. The use of ultrafiltration, to remove particles of less than 0.1 μm is a more reliable way of pre-treatment than traditional cartridge filtration and is one of the factors that has led to a greater use of thin film composite membranes which operate at lower pressures.[2] After pre-treatment the water may then have its pH value raised, and possibly be chemically dosed to prevent scaling. Depending on the membrane material it may need to be chlorinated. The treated water is then pumped using high-pressure pumps through the reverse osmosis plant.

Post-treatment

The water produced by reverse osmosis is usually acidic with no alkalinity and is very corrosive. If there has been acid dosing it will contain high levels of CO_2. It has therefore to be treated to make it both palatable and safe, so that it does not attack piping and plumbing. It is necessary to add alkalinity and hardness and adjust the pH to ensure a stable water is produced. Typically post-treatment involves passing the water through a bed of limestone or dosing lime and CO_2. Where brackish water is being treated it may be that only a proportion of the water need be treated and the RO product can be blended with untreated water. However, some post-treatment will still be required.

Some practical aspects of reverse osmosis treatment

With operating pressures up to around 70 bar it is clear that energy costs will be a very significant proportion of the operating costs of a RO plant. For seawater power use may be up to 6–10 kWh/m^3 of product water. This is a function of the following factors:

- Temperature—this affects water viscosity and diffusion rates. Higher temperatures significantly reduce energy requirements;
- Operating pressure—for an RO plant requiring feed water at 75 bar the energy input required to each cubic metre of feed water is approximately 2.1 kW h.

- Assuming a pump/electric motor efficiency of 75%, this requires electrical power of approximately 2.7 kW h/m^3 of feed water.
- Typically it takes around 3 m^3 of feed seawater to produce 1 m^3 of product water. Thus the energy required per cubic metre of product water would be 8.2 kW h.
- At a power cost of £0.05/kWh this would be a power cost of £0.41/m^3 of product water, excluding any energy recovery. Energy recovery would typically reduce the power cost by around 25%.

Although these figures are typical of a seawater plant, they will vary dependent on the design of particular plants. The reject water, which in the example above represents two thirds of the water pumped to the plant, will have a high specific energy on leaving the RO plant. It is therefore usual to install an energy recovery plant to recover as much of this energy as is economic. There are energy recovery systems under development that promise far higher-energy recovery rates than the 25% quoted above.

A key parameter in the design of a RO plant is the conversion; this is the permeate flow as a percentage of the feed flow. Spiral wound modules only achieve around 10% whereas hollow fibre modules achieve considerably higher rates. In the light of energy use as discussed above it is clearly absurd to work with a low conversion rate. Thus it is normal to use the reject water from the first set of modules as the feed to a second set, and to use the reject flow from the second set as the feed to a third set. Thus by employing a series of treatment stages a higher conversion rate can be achieved. A limit on conversion is the acceptable salt concentration in the product water and the solubility of salts in the reject water. It can be seen from equation (10.1) above that the salt flow across the membrane is proportional to the difference in concentration across the membrane. The salt concentration in the feed water increases as the water moves down the series of modules, and the salt concentration in the product water similarly increases. Also as the concentration of salt increases in the feed water there is an increasing risk of precipitation of dissolved salts. The overall conversion rate for an RO plant typically varies between around 25% for older seawater RO plants and up to over 80% for a plant treating brackish water.

Conversion rates impact not only power cost but also pre-treatment cost, which becomes very significant for seawater plants with a low conversion rate.

The cost of water produced from RO plants is dependent on several variables including:

- Water source—seawater plants cost more to build and operate than plants using brackish water.
- The size of the plant—this affects both the capital cost per unit of capacity and the operational cost. The size of a RO plant is increased largely by adding more modules and thus above a certain size the capital cost is approximately proportional to capacity. However, the cost of ancillary works such as the seawater intake are often significant and are less dependent on plant size than the RO plant itself. Similarly, some elements of operational costs such as power, membrane replacement costs, and chemical costs, are proportional to throughput or capacity, but labour costs will be proportionally higher for a small plant.
- Electricity costs—clearly a major component of the direct operating costs.
- Financing costs—the costs of financing the plant are typically of the order of £0.2/m^3 for a seawater plant with high utilization. As utilization drops the cost would increase. For brackish water the financing costs will be lower. Despite this a RO plant can be very effective at economically meeting peak demands if the costs of providing other water resources are high—for example, in tourist areas with dry summers which experience high peak demands in the tourist season.

It is quite clear that the quantity of water produced by reverse osmosis will increase rapidly over the first decades of the 21st century. The cost of water will drop as membranes improve, becoming cheaper and more efficient. However, seawater RO will always remain relatively expensive.

Nanofiltration

Nanofiltration is a form of reverse osmosis. It has developed as improvements in membrane materials have allowed membranes to be optimized for removal of larger molecules and ions. It operates at lower pressures, partly because the membranes let through most monovalent ions, leading to lower osmotic pressure, and mainly because the membranes themselves are more permeable, with a much lower head loss. Nanofiltration is not used for the production of drinking water from brackish water or seawater but is used to soften water and for nitrate removal. It can also be used to treat waters to remove organic matter that would react with chlorine to form THMs and for colour removal.[3]

Microfiltration and ultrafiltration

Ultrafiltration is a filtration process using very fine pores such that as well as all suspended matter, large organic molecules are removed, with the cut-off point dependent on the characteristics of the membrane. The process also removes all micro-organisms save possibly for smaller viruses, which will only be removed by membranes with a small pore size. Microfiltration has a larger pore size than ultrafiltration.

Most ultrafiltration membranes use hollow fibres. They normally comprise a thin skin of membrane on a supporting structure. They may have the membrane on either the outside of the fibre, in which case they will filter from the outside to the inside of the tube, or on the inside, in which case they will filter from inside the fibre to the outside. The membranes operate at far lower pressures than RO membranes, typically between 2 and 5 bar.

Ultrafiltration is a cross-flow filtration process. In a conventional filter flow is perpendicular to the filtering medium and all flow passes through the medium. In a cross-flow process the main flow is parallel to the filtering medium, with only a proportion of the flow passing through the medium. Thus rejected material tends to be carried with the reject water, being continuously removed from the filter module. This process is easier to visualize where the flow of feed water is in the centre of a hollow fibre. However, some ultrafiltration plants operate at 100% recovery rates, with all the feed water being recovered as permeate. Where this occurs the cross-flow effect is lost at the end of the process and filtered material accumulates. The need for flushing and cleaning is greater where this is the case.

As the membranes operate at lower pressures a wider choice of material is available for the membrane, and the use of more resistant materials assists in cleaning. As well as flushing and backwashing, ultrafiltration membranes are cleaned using chemicals to dissolve material attached to the membrane.

In the UK, ultrafiltration is now commonly used to provide treatment of high-quality groundwaters for *Cryptosporidium* removal, often as the only treatment apart from disinfection.

Microfiltration may use either hollow fibres or, at the coarser end, some form of filtration mat.

THE FUTURE

There seems no doubt that membrane processes will increasingly be used in the future. The reasons for the increasing popularity of these processes include:

- an increasing need to provide potable water from saline or brackish water (using reverse osmosis);
- their ability to soften water and remove nitrates (reverse osmosis and nanofiltration);
- their use for single-stage treatment of soft upland waters containing colour, iron, manganese, and micro-organisms (nanofiltration and ultrafiltration);
- their ability to effectively remove protozoan oocysts (ultrafiltration).

In all the above examples the membrane process is one of the options available and the decision as to which process to use will depend on local circumstances and, at least in part, on personal preferences.

On the other hand reverse osmosis is an energy-intensive process with correspondingly high CO_2 emissions and this may become an important factor as carbon taxes become significant. Additionally, disposal of the wastes from membrane plants, particularly RO plants, is difficult and this may limit their adoption in some locations.

11: Other processes

INTRODUCTION

This chapter considers briefly a number of processes, most of which have been developed to comply with the prescriptive water-quality standards that have come into force over the past 20 years. Several of the processes notably ozonation and GAC adsorption are still being developed. One thing that the processes considered have in common is that it would be possible to write entire books on them, and this chapter only gives a broad overview of the processes and introduces some of the key concepts. The processes considered are:

- adsorption using activated carbon—initially used in the 1970s to treat for taste and odour but adopted on a large scale in the 1980s in particular for pesticide treatment;
- ozonation—used for many years in continental Europe for disinfection but developed in the 1980s for treatment of organic chemicals, notably pesticides, and for disinfection in treatment works where chlorination would lead to failures of the THMs standards;
- nitrate removal—to meet the nitrate standard;
- air stripping—to remove volatile organic chemicals;
- chemical dosing—to reduce the corrosiveness of water in distribution systems, particularly with respect to lead;
- water softening—the odd man out in that it has been used for many years and is now little used for new plants;
- arsenic removal; and
- chemical dosing for lead treatment.

ACTIVATED CARBON ADSORPTION

Activated carbon is manufactured by heating carbonaceous material, normally coal, wood or coconut shells, under controlled conditions. Initially the material is heated in the absence of air to drive off all volatile substances. It is then activated at high temperatures by further heating it to a high temperature and allowing it to react either with steam, air, and CO_2 or with phosphoric acid. This removes the volatile matter within the pores of the carbon and produces a material with an open porous structure with a high surface area per unit weight of material.

Properties of activated carbon

There are two broad classes of activated carbon: gas adsorbent carbon, which has predominantly micropores which allow the entry of gases but permit only limited entry of liquids; and liquid-phase carbons which have a wide distribution of pore sizes and which allow liquid to penetrate the carbon.

Activated carbon comes in two main forms for water treatment: powdered (PAC), which is a finely ground material, and granular (GAC) which comes in a range of grain sizes, typically 0.5–1.5 mm. In water treatment, activated carbon is used to remove dissolved organic chemicals, either by dosing PAC to the water or by passing the water through a bed of GAC.

The properties of activated carbon are defined by several parameters. Table 11.1 lists some of the key properties of activated carbon and summarizes their significance.[1]

There is however no way to predict from the general data on carbons how an activated carbon will perform with a particular chemical or mix of chemicals without carrying out trials. These trials would normally cover several different types of carbon. However, manufacturers now have extensive data on the performance of their carbons and in practice they will often be able to advise on the treatment required. Factors affecting adsorption are discussed at some length by Benefield *et al.*[1] These include:

- characteristics of the carbon—the particle size affects the rate of adsorption, but the capacity to adsorb a particular chemical is a function of area and pore size;
- adsorbate solubility—in general the more soluble a chemical is the more difficult it is to remove, but there are exceptions;
- pH—which has a strong influence through its affect on ionization of chemicals. Organic acids are better adsorbed at a low pH, organic bases at high pH;

Table 11.1. Parameters used to characterize activated carbons

Property	Why important
Particle size	Based on standard sieve sizes. The finer the material the higher the rate of adsorption. In fixed beds finer material will have a higher head loss
Specific surface area (m^2/g)	The higher the specific area the greater the adsorptive capacity of a given weight of carbon. Sometimes referred to as the 'BET Surface Area' after the method used to calculate it
Pore volume (ml/g)	The total volume of pores in the carbon
Apparent density (kg/m^3)	The apparent (or bulk) density of the material
Specific gravity	The specific gravity of the material
Iodine number	A measure of the quantity of iodine adsorbed following a standard test. Relates to the volume of the material in the pore size range of $10–28\,\text{Å}$ and indicative of the suitability of the carbon to adsorb low-molecular-weight organic compounds
Molasses number	Similar to iodine number but measures pores greater than $28\,\text{Å}$ and indicative of ability to adsorb high-molecular-weight substances

- size of molecule—generally larger molecules are adsorbed better;
- temperature—adsorption occurs at a higher rate at high temperatures but the capacity of a carbon is reduced as temperature increases; and
- the concentration of the chemical to be removed—there will be an equilibrium established between the chemical in solution and the chemical adsorbed onto the carbon.

Adsorption isotherms

The equilibrium relationship between the amount of substance adsorbed and that remaining in solution is defined for a given set of conditions by an equation known as an adsorption isotherm. There are several forms of isotherm, depending on the theory used to model the adsorption. Two of the more common forms are the Langmuir and Freundlich isotherms. The Langmuir isotherm is based on an equation for adsorption:

$$x/m = abc/(1 + ac). \tag{11.1}$$

Where:

x = weight of material adsorbed
m = weight of adsorbent
c = concentration of material remaining in solution once equilibrium has been reached
a and b are constants.

Equation (11.1) can be rewritten as:

$$1/(x/m) = 1/b + 1/abc.$$

Thus for a system which obeys the Langmuir theory, $1/(x/m)$ plotted against $1/c$ will give a straight line with a slope of $1/ab$ and an intercept of $1/b$ when $1/c$ is zero. The values to be plotted are derived by batch testing different concentrations of the chemical of interest. Figure 11.1 is a plot of a Langmuir equation.

The Freundlich isotherm assumes that the equation for adsorption is:

$$x/m = Kc^{1/n} \tag{11.2}$$

where 'K' and 'n' are constants. Plotting 'x/m' against 'c' on graph paper will give a straight line with a slope of $1/n$ and an intercept of K where $c = 1$.

The uses of isotherms

Isotherms enable the maximum ability of a carbon to adsorb a chemical at a given equilibrium concentration to be established. This is used to calculate the theoretical quantity of substance that could be removed by a carbon while ensuring that the concentration in the treated water did not exceed a given value. However, in practice such plots cannot be directly used in water treatment for several reasons:

• there may be interference between the substance of concern and other substances present in the water which will compete for adsorption sites within the carbon;

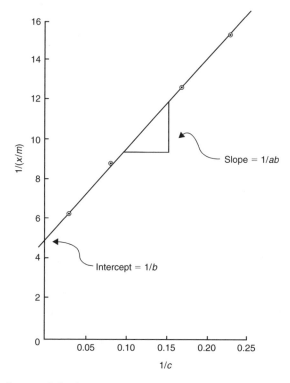

Fig. 11.1. Langmuir isotherm

- the isotherm represents the equilibrium situation. In water treatment this may be valid where PAC is dosed but where a fixed bed is used due to the kinetics of the adsorption process it would not be practicable to have sufficiently long contact time for equilibrium to be attained; and
- the data is derived from batch tests but a continuous-flow system will react differently—initially all the substance will be removed, and then the concentration in the treated water will increase steadily with time until there is no removal.

However, isotherms do permit the comparison of different isotherms. Consider Fig. 11.2 which shows idealized isotherm plots for three carbons. It can be seen that carbon 'A' has the highest adsorptive capacity at all concentrations over the range considered.

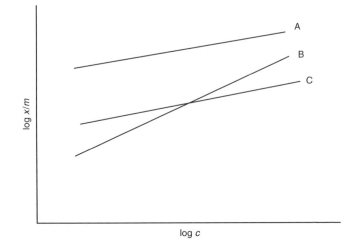

Fig. 11.2. Typical isotherm plots for different carbons

This suggests it to be the best for removal of the substance considered. On the other hand if the choice had to be between carbons 'B' and 'C', the decision would depend on the concentration of interest. Carbon 'C' would be preferred at lower concentrations, and carbon 'B' at higher concentrations.

Design of activated carbon adsorption systems

There are three possible systems that would be possible to use in water treatment:

- dosing of PAC followed by subsequent removal in a clarification stage;
- a continuous process passing water through a fixed bed; or
- a batch process using contact vessels on a fill and draw basis.

In practice the first two are used and the third is not.

PAC dosing is used where there is an intermittent problem of taste or odour or organic pollution, or as a temporary measure until a fixed-bed reactor can be installed. Drawbacks of PAC dosing include that: PAC is a very messy substance to handle; dosing has to be higher than might be needed in order to cater for variations in concentration of the substances being removed; the PAC cannot be recovered for regeneration and re-use; and there is an increase in the quantity of waste solids produced. It is fairly common for plants to have a PAC dosing facility for intermittent seasonal or emergency use but there are very few plants that dose PAC continuously. Where there is a continuing need for adsorption GAC is either placed as a media in rapid gravity or slow sand filters, or else is used in purpose built contactors. A key design parameter in GAC absorbers is the empty-bed contact time (EBCT). This is equal to the volume occupied by the GAC bed divided by the flow rate through the bed; it is normally expressed in minutes. For taste and odour removal an EBCT of around 5 min will normally be effective but for removal of pesticides and other organic chemicals a time of 15 to 20 min is normally required.

A rapid gravity filter typically has a maximum hydraulic loading of at least $6 \, m^3/m^2h$. Thus for an EBCT of 15 min a bed depth of 1.5 m is required. This is impracticably high and generally rapid gravity filters using GAC media are normally limited to use for taste and odour removal, with dedicated contactors if removal of other chemicals is required. A GAC layer 200 mm thick in a slow sand filter with a loading rate of $0.6 \, m^3/m^2h$ would have a contact time of 20 min, which is high enough for most purposes.

Figure 11.3 illustrates the theoretical performance of fixed-bed reactors in removing substances by adsorption. Figure 11.3(a) shows the reactor filled with new adsorbent, with a flow Q passing down through the reactor. The substance being treated is removed in the adsorption zone. The size of this zone depends on the velocity of the water being treated through the reactor and the rate of removal, at a low velocity or a high rate of removal this zone will be shallow. Figures 11.3(b–d) show the concentration profile of the water as it passes through the absorber. Figure 11.3(c) shows a volume of V_1 having passed through the bed and the contaminant having just reached the outlet to the absorber with breakthrough beginning. If C_2 is the maximum allowable concentration of the contaminant in the treated water, then V_2 represents the maximum volume of water that can be treated before the carbon requires to be replaced. Figures 11.3(d and e) show the GAC becoming exhausted; 11.3(e) shows the bed saturated and the treated water concentration of the pollutant the same as the treated.

An important point from Fig. 11.3 is that it is assumed that the fixed bed is in fact fixed, with the pollutant concentration given by

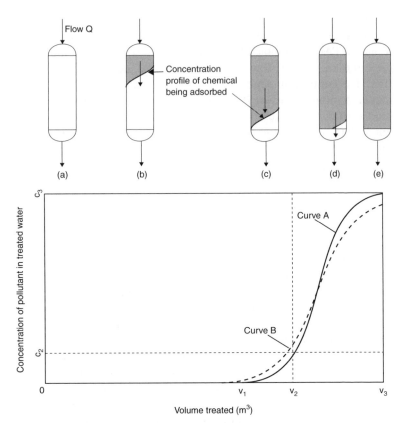

Fig. 11.3. Showing the breakthrough of pollutant in an adsorber column

curve A. In reality it is necessary to regularly backwash rapid gravity filters and it is normal to clean most contactors periodically by backwashing them using water or air and water. This however can give problems if it then results in the GAC within the filter becoming mixed. Mixing of the media will move some GAC from the top of the bed to a lower position. If the carbon that is moved was in equilibrium with the concentration in the feed water of the substance being removed, moving it to a position where the concentration is lower due to adsorption higher in the bed will result in release of the adsorbed chemical. This will lead to the concentration of contaminant following a curve similar to curve B, with the shape dependent on the degree of mixing. The quantity of contaminant retained within the GAC when fully exhausted will be the same in both cases, but in the mixed case the GAC would normally require to be replaced sooner. This problem can be overcome by ensuring that backwashing of filters/absorbers is carefully controlled to ensure full fluidization of the bed. This ensures that after backwashing the larger particles remain at the bottom of the bed and the lighter particles remain near the top of the filter. An alternative approach would be to provide two or more absorbers in series.

Practical aspects

Where adsorption is required in a plant using rapid gravity filters, it is normal now to provide GAC absorbers as a dedicated adsorption treatment stage. Where this is affordable and practical, separate adsorption is the best solution from a process point of view as it allows the adsorption process to be optimized, without the constraints imposed by the need for solids removal. Indeed some works which had rapid gravity filters modified to use GAC media have subsequently been converted back to sand filters, with separate GAC absorbers constructed. Absorbers are normally large closed vessels. They require facilities for periodic cleaning to remove solids trapped within the bed.

Chemically GAC is a strong reducing agent; it therefore reacts with chlorine or other disinfectants that may be present in the water. As a consequence there can be problems with controlling biological growth within the absorbers. Where the GAC is present in rapid gravity filters regular backwashing tends to limit problems. Where the GAC is present in dedicated absorbers it may be necessary to backwash to remove biological matter growing within the filter. An alternative approach is to accept the biological growth within the absorber and provide effective disinfection afterwards, however there can be problems associated with poor disinfection of lumps of biological material shed from the absorbers. There are some advantages to having biological growth within GAC absorbers in that the organisms may use some of the adsorbed chemicals as a substrate, prolonging the life of the GAC. This is referred to as biological activated carbon.

Where slow sand filters are used the GAC is normally placed in a layer in the lower part of the sand bed. This has proved a very cost-effective way of upgrading sand filtration plants, avoiding the high capital cost of providing a dedicated adsorption stage.

Once the GAC has reached the end of its effective life it is removed and replaced with new or regenerated carbon. Used carbon can be regenerated, with some loss of material, and re-used. Provision is made in adsorbers for placing new GAC and removing used material using an automatic system, normally as a slurry.

OZONATION

Ozone is an unstable form of oxygen with the chemical formula O_3. In solution it decays to O_2, in the process producing free hydroxyl radicals. Ozone and hydroxyl radicals are the two most powerful oxidants used in water treatment. Ozone is therefore a strong disinfectant; it is also used for oxidizing many of the synthetic organic chemicals found in water from lowland rivers and the natural organic colour found in soft waters from upland rivers. Ozone is a colourless gas with a characteristic pungent smell often associated with electricity. It is sparingly soluble in water, at 20°C the solubility at one Bar partial pressure is 570 mg/l.

Ozone is believed to react in water as an oxidant both by direct oxidation of chemicals by molecular ozone and by indirect oxidation by hydroxyl free radicals produced during decomposition of ozone.[1] Oxidation by hydroxyl free radicals is the stronger process. At low pH value direct oxidation by ozone is the more important process. At high pH value, or where the formation of hydroxyl free radicals is encouraged by UV light or by dosing hydrogen peroxide, hydroxyl oxidation may predominate.

Up until the early 1990s ozone was little used in the UK although it was widely employed in France and some other European countries. However, the European standard of 0.1 μg/l for pesticides introduced in the 1980 Drinking Water Directive and the widespread use of pesticides and their presence in most lowland rivers meant that an effective and economic method of removal was required. The method generally adopted was ozonation, to oxidize many of the pesticides, combined with GAC adsorption to remove those pesticides resistant to ozonation and the by-products of ozonation.

Ozone is manufactured by passing air or oxygen through a high-voltage alternating current discharge; this is a carefully controlled corona rather than an uncontrolled spark. After passing through the electrical discharge a proportion of the oxygen is converted to ozone. The electrodes can either be plates or, more commonly, concentric tubes.

The efficiency of ozone production compared to power input is a function of several factors in particular the oxygen content of the gas and the ozone concentration after passing through the ozonizer. Systems using pure oxygen are most efficient and systems using air are least efficient; the higher the ozone content of the gas after ozonation the lower the efficiency of power use. Other factors include temperature, the voltage and frequency used, the pressure at which the system operates, and the gap between the electrodes. Ozone production decreases at higher temperatures, and as there is a large heat production due to the electrical discharge, efficient cooling is essential.

Energy requirements for ozone generation, excluding any power used in air preparation, are in the range of 10–25 kW h/kg O_3 with the lower figure relating to low concentrations of ozone in a pure oxygen feed and the higher to high concentrations of ozone in an air feed.

It is essential that the feed supply to ozonizers is free of oil and dust and is very low in water. The presence of oil can lead to explosions and water vapour leads to uncontrolled sparking and the production of nitric acid. Preparation of the feed gas to the ozonizers, commonly referred to as 'air preparation', is as important as the actual production of ozone in the generator, and requires significantly more space.

Older systems used either air or liquid oxygen and there was little to choose between them in terms of cost, with the higher efficiency of the liquid oxygen systems balanced by the cost of the liquid oxygen. Currently the most common systems use an oxygen-enriched feed produced by treating air using pressure swing absorbers (PSA). These preferentially remove nitrogen from air, producing a gas with an oxygen content of up to 80–95% oxygen, compared to approximately 20% in the atmosphere. Energy requirements for air preparation are in the range of 10–15 kW h/kg O_3. The overall power requirement for an ozone system using air as the oxygen source is of the order of 20–40 kW h/kg O_3.

After preparation the ozone-rich gas is then dissolved in the water. This is normally done in the UK using contact tanks where ozone is introduced via diffusers in the bottom of the tanks. Alternatives are to use a venturi eductor system or to introduce ozone into a turbine or static mixer. The dissolved ozone reacts with the various inorganic and organic chemicals in the water with any residual breaking down into oxygen. Factors that are taken into account in the design of ozone contact tanks include:

- ozone concentration—the equilibrium ozone concentration in water all other things being equal will vary with the ozone concentration in the feed gas;
- bubble size—smaller bubbles have a larger surface area per unit volume of gas and thus speed up the transfer of ozone to water;
- pressure—the gas transfer rate is dependent on pressure, thus deeper tanks are more efficient;
- the ozone demand of the water—where reduced iron or manganese is present in the water ozone is removed at a high rate by the oxidation of these substances, leading to higher transfer rates than where reactions are slow;
- pH—at elevated pH values ozone rapidly decays.

A well-designed and operated system will transfer around 90% of the ozone to the water. The balance is vented to the atmosphere via an ozone thermal destruction unit. Figure 11.4 is a schematic of an ozone system.

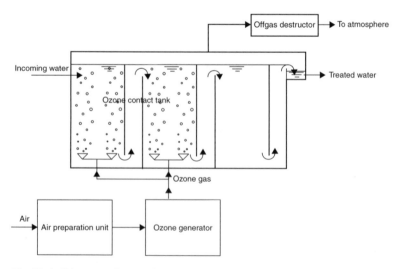

Fig. 11.4. Diagram of ozonation system

In potable-water treatment ozonation is used for several reasons:

* oxidation of organic micropollutants, notably pesticides prior to GAC adsorption;
* removal of taste and odour due to oxidation of the organic matter causing the problem;
* as a disinfectant; and
* oxidation of soluble iron and manganese.

A potential problem with ozonation is the production of bromate from water containing bromide. The 1998 Drinking Water Directive sets a bromate standard of $10\,\mu g/l$. Bromate formation can be minimized by operating at a low pH.

In practice the two most common uses of ozonation in water treatment are as a pre-treatment, prior to clarification, and as a polishing treatment after the main filtration process. The former is loosely referred to as pre-ozonation, and the latter as post-ozonation. For plants treating water from lowland rivers it is now common for both processes to be used. Pre-ozonation aims to pre-treat water: it oxidizes organic pollutants, and any iron and manganese present, kills algae, and has a beneficial effect on subsequent coagulation. It also disinfects the water. Post-ozonation oxidizes any remaining organic pollutants and disinfects the water.

When used for disinfection ozone is extremely effective against pathogenic bacteria and viruses. The effectiveness of ozone is temperature dependent, but a Ct of 1.6 will essentially kill all viruses and pathogenic bacteria. This would typically be achieved by ensuring an ozone concentration of $0.4\,mg/l$ for a contact time of 4 min. In practice where ozone is used for disinfection a typical arrangement would be to have two or three cells in series, each of around 4 min detention time. Sufficient ozone would be dosed in the first cell to give the required residual, and this would be maintained in the latter two cells. The dose required will depend on the nature and quality of the raw water. The need to limit bromate formation may impose limitations on the acceptable dose of ozone and on pH value.

When used for treatment of organic contaminants the required Ct may be higher than that required for disinfection. While the actual design criteria vary, typical criteria for treatment of lowland rivers are:

* pre-ozonation—ozone dose of around $2\,mg/l$ with the capacity to dose up to $5\,mg/l$ and a contact time of 10 min;

- post-ozonation—ozone dose of around 1 mg/l with the capacity to dose up to 3 mg/l and a contact time of 10 min.

The effectiveness of ozone to oxidize organic chemicals can be enhanced by combining ozonation with dosing of hydrogen peroxide or the use of UV light. As noted earlier this leads to increased formation of hydroxyl free radicals. In practice it has generally not been found necessary to use these enhanced forms of ozonation but they may well be required in the future to meet future standards for chemicals resistant to simple ozonation. Enhanced ozonation will permit lower ozone doses and may reduce the formation of bromate. However it does lead to control problems due to difficulties of measuring ozone concentration when free radicals are present.

ION EXCHANGE

Ion exchange involves removing unwanted ions from water and replacing them with other ions, changing the composition of the water. This is done by passing water through a bed of insoluble material made of a synthetic resin. The resins used are designed such that they will remove either cations or anions from the water passing through the bed, replacing the ions removed from the water with ions of the same charge from within the resin. The main uses of ion exchange in potable-water treatment are to remove nitrate or to soften water.

This section will touch on three aspects of ion exchange:

- classification of resin used;
- the factors affecting ion removal; and
- uses of ion exchange in water treatment.

Classification of resins used

The basic classification of resins is into cation-exchange resins (which contain exchangeable cations) and anion-exchange resins (which contain exchangeable anions). These can then be sub-divided further depending on their affinity for cations or anions. In water treatment the resins of most interest are strong-acid exchange resins, which are the cation-exchange resins used in softening, and selective strong-base resins, which are the anion-exchange resins used in nitrate removal. In water treatment both of these are regenerated with sodium chloride solution.

Regeneration

In softening, sodium ions within the resin are released into the water and replaced by cations from the water being treated. Regeneration involves introducing sodium ions into the resin, displacing the ions removed from the water. In nitrate removal, chloride ions within the resin are released into the water and replaced by anions (predominantly nitrate) from the water being treated. Regeneration involves introducing chloride ions into the resin, displacing the ions removed from the water. Thus in both cases the regenerant used is a concentrated solution of sodium chloride. Regeneration involves introducing the regenerant solution into the bed and ensuring a contact time adequate for the required degree of regeneration. After regeneration rinsing is required. Regeneration is considered further below.

Factors affecting ion removal

A resin does not remove all anions or cations equally. Rather it will preferentially remove some ions as opposed to others, and as a resin becomes exhausted there may be release of ions already removed and their replacement by other preferred ions. In weak solutions the main factors are valency, with ions of higher valency being preferred. However other factors also apply, notably the size of the ion. Size is important because ion exchange is not simply a surface phenomenon, but involves diffusion into the resin itself. Thus it is possible to optimize resins for the removal of specific anions or cations. Important physical factors include contact time, temperature, and the cleanliness and stability of the water being treated. As the ion-exchange resin is in a bed it is important that there is little or no deposition of material within the bed.

A unit weight of a particular resin will have a fixed capacity for a particular ion. This is normally expressed either as the total capacity or the operating capacity. The total capacity represents the theoretical capacity under equilibrium conditions of a resin to remove a specific ion. In practice this is of little interest and the operating capacity is far more important. This represents the capacity of the resin to remove a specific ion under the actual conditions under which the resin will be operating. Additional factors affecting this are: the ionic composition of the water being treated, as there will be competition between the ion of interest and other ions in the water; the acceptable concentration of the ion of interest in the treated water, which will define the point at which the resin will need to be regenerated; and the effectiveness of regeneration, which may not be 100% effective.

Uses of ion exchange in potable-water treatment

As noted above the main uses of ion exchange in water treatment are for softening and for nitrate removal. Softening involves removing bivalent metal cations, and nitrate removal involves removing nitrate anions. There is an important difference between the two processes. In softening it is desired to remove bivalent metal ions. Such ions will typically comprise over 80% of the cations in water. In such a water, cation replacement would largely be removing the ions it was required to remove. (Actually sodium need not be considered as no sodium is removed.) Thus ion exchange is very efficient for softening.

Nitrate removal is however different. The proportion of nitrate anions present is likely to be less than 50% of the bivalent anions. Thus a non-selective resin would remove mainly ions that were not nitrate, and would represent inefficient treatment. This has led to the development of resins that remove nitrate preferentially, greatly improving efficiency.

The key components of an ion-exchange system are shown schematically on Fig. 11.5; these are:

- the ion-exchange beds. These are normally pressure vessels similar to pressure filters containing the resin. There will normally be several vessels to allow treatment to proceed uninterrupted while one bed is being regenerated;
- a regenerant holding tank which contains the strong salt solution used for regenerating the resin;
- a used regenerant holding tank, containing used regenerant and rinse water prior to disposal off-site or to a sewer;
- the appropriate valving and transfer pumps.

Ion-exchange plants are often referred to as co-current or counter-current. This refers to the directions of normal operation and regeneration. A plant where water flows downwards during normal operation is a counter-current plant if regeneration is carried out with an upward flow of regenerant; if regeneration involved downward flow the plant would be a co-current plant.

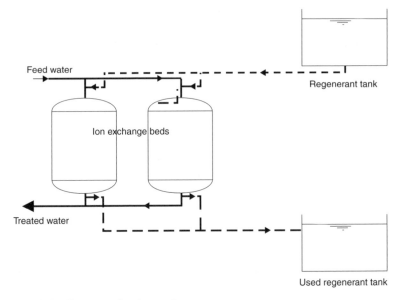

Fig. 11.5. Elements of an ion exchange system

The significance of this is that regeneration will be most complete and effective in those parts of the bed which are treated with the freshest regenerant solution. Thus in a counter-current system in which normal flow is downwards and regeneration is upwards, the part of the bed that will be most effectively regenerated is the bottom of the bed. This means that there should be no breakthrough of the ion being removed until the bed is fully exhausted. If regeneration was co-current the bottom of the bed would be the area least effectively regenerated. There will then tend to be some leakage of ions from this area of the bed during normal operation. In practice this is of little significance in potable-water treatment, although it is important in demineralization plants where counter-current regeneration is preferable.

WATER SOFTENING

Introduction

Water chemistry is a very complex and specialized subject. This short section touches on some aspects of water softening and water stabilization, and is no more than a very simplified introduction to water softening. For more detail, reference should be made to an appropriate specialized textbook (for example, Ref. 1). In this section only calcium and magnesium are considered which is reasonable for most waters.

Hard water is water that requires the use of a large amount of soap to produce lather. Soap is normally the sodium salt of a fatty acid. When soap is added to water containing polyvalent metal ions the salt formed by the polyvalent metal ions and the fatty acid precipitates, forming a scum. Only when all the polyvalent metal ions have precipitated will a lather form. Thus hardness is caused by dissolved polyvalent metal ions. In practice it is only bivalent metal ions that occur to a significant extent in normal water, notably calcium and magnesium. Waters that are high in bivalent ions can also cause scaling problems in hot-water systems.

Hardness is defined by the total concentration of bivalent metal ions present—but expressed as mg $CaCO_3$/l. This is the equivalent calcium carbonate concentration that would have the same effect as the bivalent metal ions actually present. Calcium carbonate has an equivalent weight of 50, and thus the equivalent hardness due to a concentration M of a bivalent metal ion is:

Hardness as $CaCO_3 = M$ (in mg/l) $\times$ 50/Equivalent Wt of M.

The total hardness is calculated by summing the results of the above calculation for all bivalent metal ions present. It can also be measured directly by titration.

Hardness is conveniently classified into carbonate and non-carbonate hardness. Where the metal ions are associated with bicarbonate, boiling will result in the bicarbonate being converted to carbonate and precipitating as calcium or magnesium carbonate. (It should be noted that magnesium carbonate is soluble in cold water but is less soluble in boiling water and thus is largely removed by boiling.) Where the metal ions are associated with chloride, sulfate, or other anions, with no bicarbonate present, boiling will not precipitate the bivalent metal. For this reason carbonate hardness is also referred to as temporary hardness, and non-carbonate hardness as permanent hardness. The most common cause of carbonate hardness is water passing through chalk or limestone aquifers, with non-carbonate hardness often associated with water passing through clay soils containing sulfates.

There are no hard and fast standards for hardness but water with more than 200 mg $CaCO_3/l$ is normally considered to be hard. Water softening involves the removal of bivalent metal ions. It is popular with consumers because it reduces scum formation and scaling of kettles, and it is needed for hot-water heaters and boilers to reduce or eliminate scale formation. For industrial-scale boilers and for power stations it is essential to soften the water used. There are two main ways to soften water: removal of the bivalent ions by precipitation; or replacement of the bivalent ions by sodium ions using an ion-exchange process. Other less common options are electrodialysis reversal (EDR) or reverse osmosis optimized for softening. Chemical precipitation is the process normally used in potable-water treatment. The only circumstances where ion exchange or other process might be used are where water contains a high proportion of non-carbonate hardness, or where the hardness continually varies. In industrial water treatment ion exchange is more common on boiler feed waters.

Chemical precipitation of hardness

If hydroxide, as either lime or sodium hydroxide, is added to water containing carbonate hardness, the bivalent metals will precipitate as either calcium carbonate or magnesium hydroxide. Where non-carbonate hardness is present, adding lime and carbonate to the water also results in precipitation of calcium carbonate or magnesium hydroxide. It is normal to use calcium hydroxide (lime) as the source of hydroxide, and sodium carbonate (soda or soda ash) as the source of carbonate. Thus the process used to remove both carbonate and non-carbonate hardness is commonly referred to as the lime–soda method. The proportions of lime and soda required depend on the composition of the hardness.

Lime–soda method

Removal of carbonate hardness. The first part of the lime–soda method considered is the addition of lime to precipitate carbonate hardness. Carbonate hardness is associated with bicarbonate and calcium and magnesium ions present in water.

The initial reaction on adding lime is a reaction between dissolved CO_2 and the lime. Assuming that the CO_2 is dissolved as carbonic acid, this is represented as:

$$Ca(OH)_2 + H_2CO_3 \rightarrow CaCO_3 + 2H_2O \qquad (11.3)$$

This reaction does not affect the hardness but does represent a lime demand that needs to be satisfied before there is any softening. Adding additional lime results in the precipitation of calcium and magnesium. Calcium is precipitated in a single-stage reaction:

$$Ca(OH)_2 + Ca^{2+} + 2H_2CO_3^- \rightarrow CaCO_3 + 2H_2O \qquad (11.4)$$

Thus one mole of lime removes one mole of calcium carbonate.
    Magnesium is precipitated in a two-stage reaction:

$$Ca(OH)_2 + Mg^{2+} + 2H_2CO_3^- \rightarrow CaCO_3 + MgCO_3 + 2H_2O \qquad (11.5)$$

$$Ca(OH)_2 + MgCO_3 \rightarrow CaCO_3 + Mg(OH)_2 \qquad (11.6)$$

In the first stage the bicarbonate is essentially neutralized by the lime, with calcium bicarbonate precipitating. Magnesium carbonate is soluble in cold water (but is less soluble in hot water and thus is largely removed by boiling). In the second stage addition of additional lime results in the formation of insoluble magnesium hydroxide. Thus two moles of lime are required to remove one mole of magnesium carbonate.

Removal of non-carbonate hardness. The second part of the lime–soda method considered is the addition of lime and sodium carbonate to precipitate non-carbonate hardness.
    Calcium non-carbonate hardness is removed by adding sodium carbonate:

$$Ca^{2+} + 2Cl^- + 2Na^+ + CO_3^{2-} \rightarrow CaCO_3 + 2Na^+ + 2Cl^- \qquad (11.7)$$

The calcium is precipitated as calcium carbonate, with one mole of sodium bicarbonate of lime removing one mole of calcium non-carbonate.
    Magnesium non-carbonate hardness is removed by adding sodium carbonate and lime:

$$Mg^{2+} + 2Cl^- + 2Na^+ + CO_3^{2-} + Ca(OH)_2 \rightarrow$$
$$CaCO_3 + 2Na^+ + 2Cl^- + Mg(OH)_2 \qquad (11.8)$$

The magnesium is precipitated as a hydroxide, with the hydroxide supplied by lime, and the sodium carbonate required to remove the calcium added as lime. Thus one mole of lime and one mole of sodium carbonate are required to remove one mole of magnesium non-carbonate hardness.

Other points to note. It is not essential to use the lime–soda method for precipitation of hardness. Sodium hydroxide can be used to pre-cipitate carbonate hardness and magnesium non-carbonate hardness. It will also precipitate some or all of the calcium non-carbonate hardness. Depending on the chemical composition of the water, it may be necessary to add some sodium carbonate as well.

For potable-water treatment it is normally acceptable to design to soften by calcium removal alone. This avoids the dosing of sodium carbonate, and means that only calcium carbonate is precipitated. This precipitates more easily than magnesium hydroxide and is easier to dewater. For industrial water treatment magnesium hard-ness will normally have to be removed.

Chemical precipitation does not completely remove all calcium and magnesium hardness as both calcium carbonate and magnesium hydr-oxide are not wholly insoluble. Water softened by chemical precipita-tion is unlikely to have a hardness significantly below 50 mg $CaCO_3$/l.

In practice, where a potable water is to be softened the flow will be divided into two, with one stream being softened and the other by-passing the softening process; with the flows combined down-stream of softening. The proportions of softened and unsoftened flows depend on the hardness of the raw water and the desired degree of softening required.

Dosing rates are usually higher than those required in order to speed up the reaction rate. Where part of the flow is softened this can lead to further precipitation of hardness after the softened flow is recombined with unsoftened flow, and additional chemical dosing may be required to control this.

An example of a calculation of chemical use in softening is given in Appendix 1.

Application of processes

Traditionally softening was carried out in large hopper-bottomed clarifiers designed by companies specializing in the design and construction of water-softening plants. Where chalk or limestone borehole waters were being softened the only purpose of the clarifiers was for the removal of precipitated hardness. Such clarifiers removing calcium hardness were operated at a loading rate of around 4 m/h. The sludge produced from softeners consists of fine calcium carbonate and can be difficult to thicken, dewater and dispose of.

Where clarification is required to remove suspended solids it can be combined with softening. However softening generates a high pH value, of around 10.5, and this places constraints on the coagulant that can be used. It is normal to use a ferric coagulant which works well at high pH.

Nowadays where softening is required, pellet reactors are normally used. These consist of vertical cylindrical cylinders with a conical bottom. The water to be softened and lime are injected into the bottom of the conical section. The reactor contains a bed of single size sand. The flow of water liquidizes the bed. As the softening reactions take place the calcium carbonate and magnesium hydroxide precipitate onto the grains of sand, which act as seeds. The sand particles become coated with precipitated material and grow in size. Once the reactor has stabilized it is necessary to regularly remove the heavier particles and replace them with sand.

The reactors operate at high-loading rates, above 50 m/h. The great advantages of using pellet reactors are the high-loading rates and the nature of the waste material; this consists of hard pellets of around 2 mm diameter. These require no dewatering and are easy to handle and dispose of. Pellet reactors work excellently when the hardness is predominantly calcium hardness but are not to be recommended where magnesium is present to any extent (>25 mg/l) because magnesium does not react and precipitate as quickly as calcium to form pellets. The design of these reactors is a specialized task and is normally done by the manufacturer/supplier of the plant.

Post-softening treatment

The water leaving a softening plant will be supersaturated with calcium carbonate and will have a high pH. It will therefore tend to continue to precipitate calcium carbonate unless the pH value is lowered. The easiest way of doing this is to dose a strong acid, such as sulfuric or hydrochloric acid. This however may produce a corrosive water and it is often better to lower the pH value by injecting CO_2, a process often referred to as recarbonation. An alternative is to dose a proprietary chemical such as Calgon.

Softening by ion exchange

Ion-exchange treatment has been discussed above. Where ion exchange is used in softening the resin is a cation-exchange resin. The water to be softened is passed through the bed of resin. Sodium is released from the resin and replaced by calcium, magnesium, and other metals present in the water. Thus the treated water will contain no bivalent ions and will have effectively zero hardness. When the resin is exhausted it is regenerated by a strong solution of sodium chloride.

The water produced by ion-exchange softening will have a high sodium content. However for potable treatment only a proportion of the water would be softened, and the softened water would then be mixed with the unsoftened water, lowering sodium levels.

The wastewater produced by ion-exchange softening plants is a strong saline solution, also containing the metals removed in the softening process. This can normally only be disposed of to a sewer, where it is diluted with sewage to lower salinity to an acceptable value. Ion exchange has been used in the UK for potable-water softening but it seems unlikely that many new plants will be constructed in the future because of problems of waste disposal.

REMOVAL OF IRON AND MANGANESE

Both manganese and iron can exist in different oxidation states. Iron exists as either +2 or +3. Within the normal pH range of interest in water treatment Fe^{2+} is soluble and Fe^{3+} is insoluble. Manganese exists in a range of oxidation states but effectively two states are of interest, Mn^{2+} is soluble and Mn^{4+} is insoluble.

Well-aerated surface waters from rivers or reservoirs will normally only contain low levels of iron and manganese. This is because the iron and manganese will be oxidized and present as insoluble forms and will tend to settle out of the water. Where they are present they are easily removed during conventional treatment.

Groundwater from confined aquifers and water from the lower anoxic zones of reservoirs often contains the reduced soluble forms of iron and manganese. Water from reservoirs may also contain iron or manganese complexed with organic material. This section considers some aspects of the removal of dissolved iron and manganese. References 1 and 3 cover in more detail the theoretical and practical aspects of iron and manganese removal respectively.

Where inorganic iron and manganese are present, the treatment required is oxidation of the iron and manganese, which will cause them to precipitate, followed by removal of the precipitated material. Oxidizing agents commonly used are oxygen and chlorine; other possible oxidants are ozone, chlorine dioxide, and potassium permanganate.

Iron is relatively easy to oxidize. The requirements are a pH of 7.0 or above and the presence of dissolved oxygen. The process proceeds rapidly but can be further speeded by chlorination. The rate of oxidation is affected by both temperature and pH, proceeding faster at high pH values and temperatures. Figure 11.6[1] illustrates the effect

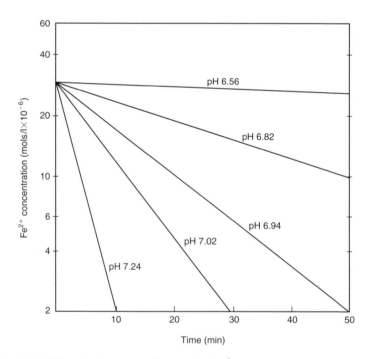

Fig. 11.6. Effect of pH on rate of iron oxidation[1]

of pH on reaction rate. In practice the pH value is adjusted either by chemical dosing or, if there is significant dissolved CO_2, by aeration. Aeration may itself need to be followed by dosing to stabilize the water if it leads to precipitation of calcium carbonate.

Manganese has a reputation for being more difficult to remove. This is because it reacts more slowly than iron and may pass through treatment only to precipitate out later. Manganese oxidation by oxygen requires a pH value of 7.5–10, and proceeds only slowly below a pH value of 9.5. Figure 11.7[1] illustrates the effect of pH on reaction rate. Where manganese is present it is important to dose a chemical oxidant, normally chlorine, to speed oxidation. However the oxidation of manganese is complicated by catalytic effects. If water containing dissolved manganese is passed over manganese dioxide there is normally a strong catalytic effect and the manganese oxidizes and precipitates onto the manganese dioxide more rapidly than would otherwise be expected.

Where there are significant concentrations of both iron and manganese it is fairly common to provide two-stage treatment, with iron oxidation and filtration at a pH value of around 7.0, followed by

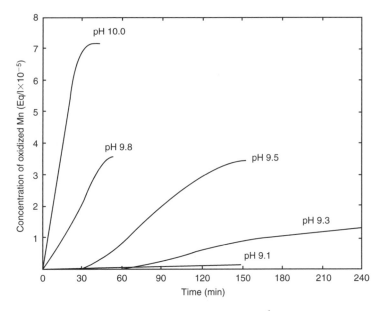

Fig. 11.7. Effect of pH on rate of manganese oxidation[1]

manganese oxidation and filtration at a higher pH value. The iron filters will normally be loaded at a rate of 6–9 m/h and the manganese filters at 10 m/h or higher. This is a reliable way of treatment but is expensive and arguably is not normally necessary as in most cases single-stage removal should be possible. However the need to guarantee compliance with water-quality standards with minimum operator input makes two-stage removal attractive to water companies.

Where iron or manganese are present complexed with organic matter, as they can be in water drawn from the lower anoxic zones of reservoirs, oxidation and precipitation is not possible except at elevated pH. Where this is the case coagulation followed by clarification and filtration is normally effective.

Other options

Ion exchange can be used for iron and manganese removal; in this case it is important not to aerate the water prior to treatment in order to avoid iron and manganese precipitating out of solution with the ion-exchange bed. Iron and manganese can both be removed biologically, but the optimum conditions required for each are different, precluding single-stage biological removal of both metals. In practice biological removal of manganese is unusual whereas there are many plants which treat groundwater to remove iron.[3]

Biological iron removal involves passing water through filters, normally pressure filters, with careful control over dissolved oxygen. The key requirement is a low dissolved oxygen level, of around 1 mg/l, in the feed water, which should not be chlorinated. High filter loadings can be used, and because the iron is present in a dense form, the filters can treat waters with a higher iron concentration than could be treated by a conventional oxidation/filtration process. After treatment the water needs to be aerated and disinfected. The process is not suitable for all groundwaters.

NITRATE REMOVAL

There has been a steady increase in nitrate levels in many groundwater sources. This together with the nitrate standard of 10 mg/l as nitrogen, introduced in the first EU Drinking Water Directive has led to a number of sources being unsuitable for water supply without blending with lower nitrate water, or nitrate removal.

Where a source does have high nitrate levels the first preference is blending. Where this is impractical or expensive then nitrate removal will be required. There are two main options for nitrate removal: ion-exchange or biological treatment. It is also possible to use reverse osmosis or electrodialysis.

The simplest treatment option is ion exchange. Where this is used, a proportion of the flow passes through an ion-exchange unit using resin optimized for nitrate removal. The advantages of ion exchange are simple operation; the process is independent of temperature, can be automated, and is essentially unaffected by varying nitrate concentrations. The drawbacks are the production from regeneration of a saline waste, that may cause disposal problems, and an increase in the corrosiveness of the treated water. The use of ion exchange for nitrate removal normally requires the use of a special resin to be practicable. A conventional resin removes sulfate ions in preference to nitrate ions. Thus where both sulfate and nitrate are present both are removed, reducing the effective capacity of the resin for nitrate, and increasing the regeneration water per unit of nitrate removed. There is a further potential problem associated with the use of conventional resins where both sulfate and nitrate are present: if regeneration is delayed until the capacity of the resin has been exceeded there will be a higher nitrate concentration in the treated water than in the influent. This is associated with the displacement of nitrate previously removed by sulfate. (However in practice this should not be a problem for an automated plant.) Thus nitrate removal normally uses special resins that remove nitrate in preference to sulfate.

Biological denitrification relies on bacteria to reduce the nitrate to nitrogen gas. In theory there are two basic approaches: the process can use either autotrophic or heterotrophic organisms. Heterotrophic bacteria use an organic carbon source, such as acetic acid or ethanol to provide energy for growth. In this case the water requires to be dosed with the carbon source at a concentration that provides sufficient nutrient for the bacteria while avoiding the carry-over of nutrient into treated water. Requirements include a temperature of greater than approximately 8°C and a low dissolved oxygen concentration, in order to ensure the reduction of nitrate. The process uses attached growth with the bacteria growing on a fluidized bed of sand. The process requires careful control and cannot easily be automated. The dose of the organic carbon source has to be matched to the nitrate level. The use of autotrophic bacteria, which use inorganic carbon within the water as the source of carbon for growth, is not normally practicable as they grow much more slowly than heterotrophic bacteria and the process is troublesome.

AIR STRIPPING

Where water is contaminated with a volatile organic chemical, air stripping is often the most cost-effective method of treatment either to lower the concentration to an acceptable value, or as pre-treatment prior to GAC adsorption. Many volatile organic chemicals are poorly adsorbed by GAC whereas they are easily removed by air stripping. Examples of chemicals that can be economically removed by air stripping include trichloroethane and tetrachloroethane. Air stripping is also used for removal of CO_2.

The theory of air stripping does not lend itself to a simple summary. It is based on the fact that if the partial pressure of a volatile chemical is greater in a solution than in the air with which it is in contact, the chemical will move from water to air. Air stripping provides the optimum conditions necessary for the transfer to proceed. The rate of transfer depends on many factors the most important of which are:

- the difference in vapour pressures—the rate of transfer is proportional to the difference in vapour pressures which can be thought of as the driving force;
- the solubility of the volatile chemical (which is related to the Henry constant of the chemical; soluble gases have a low Henry constant);
- the surface area across which transfer takes place;
- the mixing within the liquid, bringing water with a higher concentration of pollutant in contact with air; and
- the ease with which the volatile chemical diffuses within the liquid.

Air stripping normally involves a tower filled with a packing material. Water passes down the tower while air is blown upwards. In an air stripping process the air entering the tower will contain none of the volatile chemical being stripped from the water. As it passes upwards the concentration of the volatile chemical increases to a maximum at the top of the tower. The opposite is true of the water; it has a maximum concentration of contaminant at the top, and a minimum concentration at the bottom. For a given system the rate of transfer of the chemical will depend on the difference in vapour pressures between the water and air phases. In practice the highest transfer rate will be at the top of the tower and will decrease as the water passes down the tower.

The process is usefully thought of as a series of transfer units in each of which the air and water vapour pressures attain equilibrium. In the top transfer unit the equilibrium is between a relatively high concentration of the chemical to be removed in both the air and water; by the time the bottom of the tower and the last transfer unit is reached the concentrations in both the air and water are low. The design of an air stripping tower is then undertaken by determining how many of these theoretical transfer units are required; this will depend on the inlet concentration and desired effluent concentration of the chemical to be removed; the airflow; and Henry's constant for the chemical at the operating temperature. Fewer transfer units are required when:

- Henry's constant is high;
- the airflow is high;
- the contaminant concentration in the feed water is low; or
- the required contaminant level in the treated water is relatively high.

(The latter three of these conditions mean that the contaminant concentration in the air leaving the stripping tower will be low, as can be demonstrated from a mass balance calculation for the air and water streams.)

The converse is that when the opposite of the above holds, more transfer units are required. In practice of course the process is not a series of discrete units but a continuous process and the number of transfer units is calculated using a logarithmic formula.

There are various forms of packing used in air-stripping towers, these are normally made of plastic. The aim is to provide the maximum surface area per unit of packing volume. One of the oldest forms is the Raschig ring, thin-walled open cylinders with a length equal to their diameter. However these are not particularly efficient and more complex shapes are common. For a given air-stripping system the key parameter of a particular form of packing is the height of a transfer unit. This and the number of transfer units defines the height of tower required.

Thus the design of an air-stripping tower needs to take account of the factors touched on above. In practice the specialist suppliers of air-stripping towers will do the process design.

Once an air-stripping tower is installed it is possible to vary the airflow to change the performance of the tower. Increasing the airflow decreases the number of transfer units required to attain a given treated-water quality, providing more effective treatment. Problems with air stripping include deposition of iron and manganese if present in the raw water and dirt and contamination from the air if no filter is provided.

Water that has passed through an air-stripping plant may need post-treatment to stabilize it.

ARSENIC TREATMENT

The 1998 European Drinking Water Directive includes an arsenic standard of 10 µg/l, reduced from 50 µg/l in the 1980 Directive. This has meant that arsenic-removal plants are having to be considered for some British water sources. Some arsenic removal may occur in treatment plants using ion exchange, activated carbon adsorption or coagulation and filtration for iron-removal plants, but it has not been necessary in the past to design plants for arsenic removal.

The processes that can be used for arsenic removal are either adsorption onto a suitable media or reverse osmosis. Unless there are other water-quality issues that require reverse osmosis treatment, adsorption will normally be preferred. The process involves passing water through a bed of adsorbent media. Possible adsorbents include:

- granular ferric hydroxide;
- activated alumina; and
- bone char.

Ion-exchange resins and activated carbon may also be used.

Activated alumina and granular ferric hydroxide are considered to be the most likely adsorbents for a dedicated arsenic-removal plant. The design issues that need to be considered are the same as for other adsorptive processes: contact time; hydraulic loading rate, absorptive capacity; and backwashing/regeneration. The quality of the feed water needs to be high, with low turbidity and solids so adsorption is appropriate only for high-quality groundwater or treated surface waters. EBCT is normally a minimum of around 5 min. The permissible hydraulic loading rate depends on the particular media used, maxima of 10–20 m/h are appropriate. However in practice the acceptable hydraulic loading rate may be determined by EBCT and physical constraints; an EBCT of 5 min and a loading rate of 20 m/h requires a media depth of 1.7 m, and this is high for pressure vessel adsorbers.

The adsorptive capacity of the media depends partly on the pH value of the water being treated. The media has a greatly increased capacity at lower pH values and a pH of 5.5–6.0 is desirable.[4] However the extra costs and complexity of pH adjustment may not be economic. Both granular ferric hydroxide and activated alumina can be regenerated by passing sodium hydroxide through the media but in practice it appears that for most sites it will be preferable to replace media when it is exhausted.

As with all adsorptive processes backwashing should fully fluidize the media to prevent mixing of media during operation.

CHEMICAL DOSING FOR LEAD CONTROL

The 1998 EU Drinking Water Directive introduces a maximum allowable concentration of 25 μg/l from the end of 2003, reducing to 10 μg/l at the end of 2013. The 25 μg/l standard will be difficult to attain for many waters, and the 10 μg/l will be impossible, without reducing lead solubility.

Where water comes into contact with lead in the presence of carbonates, the carbonate in the water reacts with lead forming one of several lead carbonates, mainly either lead carbonate ($PbCO_3$) or basic lead carbonate ($Pb_3(CO_2)_2(OH)_2$). The species formed depends on pH and alkalinity. The concentration of lead carbonate in water is a minimum at a pH value of around 9–9.5. At this pH value the lead concentration is lowest in low-alkalinity waters, with a minimum lead concentration of around 25 μg/l.[5] However, low-alkalinity waters have low buffering capacity and it is often difficult to maintain the pH of the water as it moves through the distribution system.

Where the water contains dissolved orthophosphate the concentrations of lead are much lower as lead phosphate compounds are formed and these are less soluble than lead carbonates. Dosing orthophosphate reduces minimum lead concentrations to around 5 μg/l. In a low-alkalinity water the pH range required for minimum lead concentrations is 7.5–8. As alkalinity increases a pH in the range of 7.2–7.8 is required.

Thus one approach to meeting the lead standard is to dose orthophosphate and control pH. Sufficient orthophosphate has to be dosed to maintain the necessary concentration of phosphate at the ends of the distribution system. The dose required depends on the chemical composition of the water, alkalinity, pH, and the condition of the distribution system. Typically between 0.6 and 1.8 mgP/l is dosed to achieve the phosphate level required to minimize lead concentrations throughout the network. However, it is not sufficient simply to dose phosphate but it is also necessary to ensure that the optimum pH value is attained, and also maintained, as the water passes through distribution. This has implications for treatment, requiring that chemically stable water be produced at the treatment works. This can be difficult, particularly for low-alkalinity waters.

At the time of writing it appears that this approach is the only practicable treatment option to meet the new lead standards; the alternative will be to remove lead from plumbing systems.

12: Disinfection

INTRODUCTION

As water passes through a treatment plant, the various processes remove or inactivate many of the organisms present in the raw water. The final treatment in a water-treatment plant is disinfection of the treated water. The correct meaning of 'disinfection' is under threat and the word is often incorrectly used. Disinfection is not sterilization, which implies the inactivation of all organisms; rather it is the killing of pathogenic organisms, those which cause disease. Disinfection is normally the most critical process from the point of view of supplying safe water. There are two aspects to disinfection: the first is the disinfection of the water to kill all pathogens that have passed through the various treatment stages, and the second is to apply a residual disinfectant so that the water leaving the treatment works remains safe as it passes through the distribution system to the point of use.

There are three sorts of pathogenic micro-organisms that are of greatest concern in water treatment: viruses, bacteria, and protozoa. It is not straightforward to classify micro-organisms, which is complicated by there being a continuum of organisms, rather than a number of distinct types, but it is useful to understand some of the key differences. Viruses are the smallest. They are simple organisms, consisting of a core containing nucleic acid (either RNA or DNA) surrounded by an envelope of protein. They are obligate parasites able to multiply only in other living cells. It is quite possible to argue that viruses are not living organisms at all. Viruses are responsible for diseases such as polio, AIDS, rabies, and the common cold. Some pathogenic viruses can be transmitted in water. Bacteria and protozoa are both single cell organisms. They are both normally heterotrophic, using organic matter as the source of energy. Some of the difference between bacteria and protozoa are that bacteria are smaller and lack a clear nucleus; they evolved much earlier than protozoa and tend to be less complex. Examples of bacteria of concern in water treatment include *Salmonella typhi*, which causes typhoid, and *E. coli*, which is the key indicator of faecal contamination of water. Examples of protozoa of concern include *Cryptosporidium parvum* and *Giardia lamblia*. A key factor of some of the parasitic protozoa, notably *Cryptosporidium parvum*, is that they have a complex reproductive cycle which includes the production of oocysts. These are the infective bodies that transmit the organism from host to host. The oocysts are thick-walled to protect the infective material within. The oocysts can be very resistant to chemical attack and can exist dormant for many months.

All the disinfection processes discussed in this chapter are effective against pathogenic bacteria and viruses. They may not be effective against some spore-forming bacteria, but these are not pathogenic. All disinfection processes are considered by the regulatory authorities in the UK to be ineffective against *Cryptosporidium* oocysts at acceptable or practicable dosages. Thus, such organisms have to be physically removed. In the USA, ozonation and UV disinfection under specified conditions are considered effective for inactivation of *Cryptosporidium* oocysts.

The need for a residual disinfectant to be carried into the distribution system is taken for granted in the UK, and over much of the world. However, in some countries[1] there is increasing public resistance to the use of chlorination as a residual disinfectant under normal conditions. If a distribution system is in good condition and clean, and the water entering the system is stable and low in dissolved assimilable organic carbon, then there will be little bacterial growth within the distribution system and it can be argued that disinfection is not required save where the security of the system has been compromised by a burst pipe or other event. It is quite possible that there will be increasing objections to chlorination in the future and such a view may become more widely held.

DISINFECTION DURING TREATMENT

Clearly the disinfectant used must kill the organisms of concern while not being toxic to humans. Disinfection within the treatment plant may utilize one or more of the following processes:

- chlorination;
- ozonation;
- UV disinfection;
- chlorine dioxide.

Chlorination is now mainly used only for disinfecting fully treated water or high-quality groundwaters; this is to reduce the formation of THMs. These are chemicals formed from methane (CH_4) in which three of the hydrogen atoms have been replaced by halogens (primarily chlorine). The compound of most concern is chloroform ($CHCl_3$), which is a known carcinogen. THMs are formed by reactions between halogens and organic matter in water, in particular the humic and fluvic acids found in peaty-coloured water. Commercial chlorine gas also contains some bromine and hence THMs can also contain both. In the past, it was common to pre-chlorinate water at the point of entry to the treatment works. This controlled the attached growths of algae that otherwise occur and also killed algae present in the raw water, making them easier to remove. Nowadays, it is unusual to pre-chlorinate, although it may be acceptable if an assessment of THM formation potential indicates that there will be no problems with THMs. Chlorine dioxide may be used for waters that are particularly prone to THM formation.

Ozonation has been covered in Chapter 11. It has been widely used in Europe for many years for disinfection in water-treatment plants. It is also now becoming common on large plants in the Britain and North America. In Britain, it is used primarily for oxidation of pesticides, but it also disinfects. UV disinfection may be favoured where an existing plant does not have a contact tank for chlorination.

Residual disinfection

Because chlorine-based disinfectants are persistent, the residual disinfectant is always chlorine based, and may be:

- chlorination (using either gas or hypochlorite solution);
- chloramination (using chlorine and ammonia); or
- chlorine dioxide.

In distribution systems, chlorination is the most common disinfectant, but chloramination is also widely used.

THEORY OF DISINFECTION

It is believed that chemical oxidants kill bacteria by rupturing cell membranes and destroying enzymes. There are many variables in the disinfection process, all of which affect its efficiency. The effectiveness of chemical disinfectants depends on several factors[2] summarized below.

The organisms to be killed. Chlorination can kill essentially all bacteria and viruses, although some are more resistant than others. *E. coli* is considered more resistant than most pathogenic bacteria and some viruses and this is one of the reasons why it is used as an indicator organism. However, some pathogenic viruses are believed to be more resistant to chlorination than *E. coli*, and thus there are other requirements for disinfection than simply killing *E. coli*. Chlorination and other disinfectants do not under normal circumstances kill protozoan oocysts, most notably *Cryptosporidium parvum*, which therefore have to be removed by treatment prior to disinfection.

The nature of the disinfectant. Chlorine and hypochlorite can form a number of different chemical species in water dependent on pH value and the presence of other chemicals in the water. The effectiveness of these as disinfectants varies greatly. This is discussed further below.

Concentration of the disinfectant. The effect of concentration is given by:

$$C^n T = K$$

where 'C' is the concentration of the disinfectant, 'n' is coefficient of dilution (or the order of the reaction), T is the time to achieve a given percentage kill, and 'K' is a constant. It can be seen that if 'n' is greater than 1, then concentration has more effect than contact time, whereas if 'n' is less than 1, then contact time has more effect than concentration. The values of 'n' and 'K' depend on many factors including temperature, pH, the disinfectant being used, and the organism being killed. For chlorine, 'n' is normally taken as 1.

Contact time. In theory, the kill of organisms should follow Chick's law. This states that:

$$dN/dt = -kN$$

where 'N' is the number of surviving organisms and 'k' is a rate constant for a particular disinfectant/organism combination. In other words, the rate of kill is proportional to the number of living organisms. By integration it can be shown that when 'kt' equals 1, 43.4% of the original organisms will have been destroyed. However, in practice the effect of contact time is more complex. For many organisms the rate of kill increases with time. It is postulated that this is related to the time required for the disinfectant to enter and kill organisms.

Temperature.   Increases in temperature lead to increased rates of chemical disinfection.

Other factors.   These include the presence of solids in the water that may protect organisms from the disinfectant, and the concentration of organisms in the water. Thus disinfection of turbid water can be ineffective, hence the need to disinfect water of low turbidity.

Disinfection is a complex process that is difficult to model, given the wide range of variables that affect its efficiency. In practice simple design criteria that give adequate reassurance of effective disinfection are applied; the effectiveness of disinfection is verified using microbiological indicator parameters, notably the absence of coliform and *E. coli* organisms in the treated water.

CHLORINATION

Chlorination is the most common form of disinfection. It can involve two main alternatives; the use of gaseous chlorine, which is dissolved in carrier (motive) water before being added to the water to be treated, or the use of a solution of hypochlorite, normally sodium hypochlorite.

When added to water, chlorine reacts rapidly to form hypochlorous acid and hydrogen and chloride ions (effectively dissolved hydrochloric acid):

$$Cl_2 + H_2O \rightleftharpoons HClO + H^+ + Cl^- \qquad (12.1)$$

The hypochlorous acid may then dissociate:

$$HClO \rightleftharpoons H^+ + OCl^- \qquad (12.2)$$

The equilibrium points in these reactions are pH dependent. Equation (12.1) applies at a pH value of less than approximately 4, with the proportion of gaseous chlorine dropping to zero at a pH of 4. Equation (12.2) covers a pH range of 5–10, with the hypochlorous

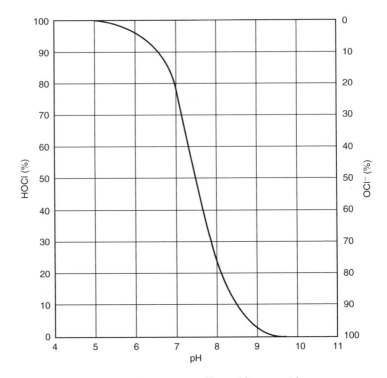

Fig. 12.1. Effect of pH on dissociation of hypochlorous acid

acid being fully dissociated at a pH of around 10. Figure 12.1 illustrates the effect of pH on hypochlorous acid dissociation.

Where a solution of sodium or calcium hypochlorite is used they dissociate as follows:

$$Ca(OCl)_2 \rightleftharpoons Ca^{2+} + 2OCl^- \qquad (12.3)$$

$$NaOCl \rightleftharpoons Na^{2+} + OCl^- \qquad (12.4)$$

However, the equilibrium in (12.2) also applies, and in addition the sodium or calcium ions will also affect the hydrogen ion concentration. Sodium and calcium hypochlorite therefore have identical reactions in water as chlorine, but the presence of the metal ions affects the concentration of hydroxide ions, resulting in an increased pH value, whereas chlorine gas lowers the pH value.

The other important set of reactions in relation to chlorination is the reactions between chlorine and ammonia. These are important firstly because they remove hypochlorous acid and secondly because the compounds formed are disinfectants and are often used to provide residual disinfection in distribution systems. Chlorine or hypochlorous acid reacts with the ammonium ion to successively replace the hydrogen atoms with chlorine:

$$NH_3 + HOCl \rightleftharpoons NH_2Cl + H_2O \tag{12.5}$$

$$NH_2Cl + HOCl \rightleftharpoons NHCl_2 + H_2O \tag{12.6}$$

$$NHCl_2 + HOCl \rightleftharpoons NCl_3 + H_2O \tag{12.7}$$

These reactions lead to the successive formation of monochloramine, dichloramine, and trichloramine. The relationship between the amounts of the three types of chloramine depends on the pH value and the NH_4 concentration of the water. As trichloramine can only form at very low pH values, the other two prevail in water treatment, dichloramine being much the more powerful bactericide. However, chlorination of water containing ammonia also leads to the production of nitrogen gas as follows:

$$2NHCl_2 + HOCl \rightarrow N_2 + 3HCl + H_2O \tag{12.8}$$

This releases nitrogen and converts hypochlorite to hydrochloric acid. There is also some production of nitrate, but this is a minor reaction.

Hypochlorous acid and the hypochlorite ion together are called 'free chlorine'. The chloramines are known as 'combined chlorine'.

Breakpoint chlorination

When chlorine or hypochlorite is added to water a succession of reactions occur. Refer to Fig. 12.2. Even the purest water tends to have slight traces of ammonia and other chemicals that react with chlorine. Thus the first reactions result in the removal of some chlorine, due to reactions with ferrous or sulfide ions for example, and the formation of monochloramine and dichloramine. This results in an increase in chlorine residual that closely follows the applied chlorine dose, with the difference due to other reactions. Once most of the ammonia has been converted to dichloramine equations (12.7) and (12.8) start to occur. Equation (12.8) results in formation

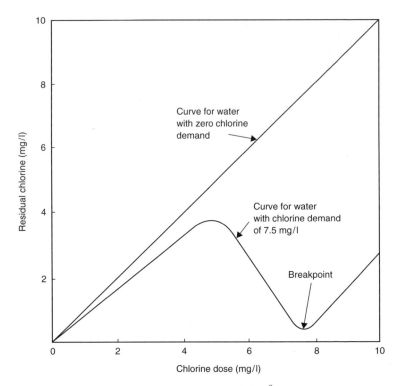

Fig. 12.2. Breakpoint chlorination curve (after White[8])

of hydrochloric acid and the chloride ion, neither of which are detectable as chlorine residual. Thus the second stage of the curve shows the residual chlorine concentration decreasing with increased chlorine dose, with this apparently anomalous effect being associated with the formation of nitrogen. Once all the dichloramine has reacted to form trichloramine or nitrogen, the chlorine residual increases as the chlorine dose increases. The point at which the chlorine residual starts to increase again is referred to as the breakpoint. Prior to the breakpoint chlorine is present predominantly as combined chlorine, after the breakpoint free chlorine predominates. Understanding this curve is critical to an appreciation of chlorination and chlorine demand.

The 'chlorine demand' of any given water is the amount of chlorine required to take the reaction to the break point. In some waters this is very low and often does not exceed or even reach 0.3 mg/l, but in other waters it may be much higher. It has to be determined by experiment.

Marginal chlorination, superchlorination, and dechlorination

The most effective form of chlorination is superchlorination. This involves dosing sufficient chlorine for there to be residual free chlorine; in other words sufficient chlorine is dosed to ensure that the breakpoint is passed. This is the method generally favoured as it is the safest method of disinfection. It is normal to dose sufficient chlorine to obtain a free residual dose of around 1.0 mg/l. However, this produces water in which the chlorine can be smelt or tasted by users, many of whom find it unpleasant. It is therefore common to allow sufficient contact time to ensure full disinfection and then to dechlorinate. This is normally done by dosing sulfur dioxide, which reduces the chlorine to chloride, lowering the free chlorine to around 0.5 mg/l.

Marginal chlorination involves dosing sufficient chlorine to give the desired combined chlorine concentration. As the chlorine is combined the disinfection is much less effective. Arguably marginal disinfection should only be used on very high-quality waters, either high-quality groundwaters or surface waters that have undergone extensive treatment including another disinfection stage. Marginal chlorination can give rise to complaints of taste and odour resulting from the reactions between chlorine and the water being chlorinated, particularly where is organic matter present in the water. For this reason many suppliers prefer superchlorination.

Chloramination

One of the problems often found with chlorination is that in order to maintain a residual throughout the distribution system, it may be necessary to apply a very high dose at the treatment works. This occurs where the distribution system is old and contains deposits with which the chlorine reacts, or in large systems with a high residence time. It is sometimes impracticable to dose at the high rate required to maintain a residual throughout the system; this could be because a high rate would give rise to complaints of chlorine taste and odour from consumers located near to the treatment works, or it might be because of water-quality issues—high chlorine levels can give rise to high THM levels in some waters. It may be that the problem can be overcome by booster chlorination at service reservoirs, or it may be possible to use chloramination. Chloramination involves dosing a controlled dose of ammonia to chlorinated water. This converts the chlorine to monochloramine. This is a weaker disinfectant than free chlorine, and for this reason it reacts less with material in the distribution system and is far more persistent. However, there are problems in that as the monochloramine decays it leads to the formation of nitrite. The 1998 EU Drinking Water Directive has set a standard for nitrite at the treatment works to ensure that there is close control of chloramination. At present chloramination is widely practiced, notably for most of London's water, but is possible that in the future it will be used less.

Chlorine dioxide

Chlorine dioxide is a compound formed by a reaction between sodium chlorite and either chlorine or hydrochloric acid. It is manufactured by mixing aqueous solutions in a small reaction vessel. At ambient temperatures undissolved chlorine dioxide would be a gas. However, as the gas is explosive one of the key aspects of making chlorine dioxide is to ensure that it remains in solution:

$$5NaClO_2 + 4HCl \rightarrow 4ClO_2 + 5NaCl + 2H_2O$$

$$2NaClO_2 + Cl_2 \rightarrow 2ClO_2 + 2NaCl$$

The attractiveness of chlorine dioxide is that while it is a strong disinfectant and oxidising agent, it does not lead to the formation of THMs. It has a similar disinfection efficiency to free chlorine and is effective over a pH range of 6–9. At a pH value higher than 10 it is ineffective.[3]

A concern with chlorine dioxide is that its production can lead to the formation of both chlorate and chlorite as by-products. Chlorine dioxide can also break down to chlorate and chlorite in the distribution system. Chlorite has been associated with affecting reproduction in laboratory animals, and chlorate can be reduced to chlorite. To minimize chlorate and chlorite formation it is important that in the manufacture of chlorine dioxide there is precise control of the proportions of the reacting chemicals; it is also necessary to maintain a low pH value. There are no standards for chlorate or chlorite in either the current EU Drinking Water-quality Standards or current UK Drinking Water-quality Regulations. However, chlorite and chlorate are controlled in the UK by a requirement that the combined concentration of chlorine dioxide, chlorite and chlorate must not exceed 0.5 mg/l as chlorine dioxide.[4] The USA sets a limit of 1 mg/l for chlorite.

REQUIRED DISINFECTANT DOSE AND CONTACT TIME

Disinfection is effectively required to kill all pathogens. In practice the design criterion will be the effective killing of *E. coli*. To do this, the two key parameters are contact time '*t*' and disinfectant dose '*C*'. It is normal to quote '*Ct*' values in the design of disinfection systems. This represents a simplification of a complex relationship but appears satisfactory provided there is a safety margin included.

Contact time required

The effectiveness of the various forms of chlorine differs significantly. There is a wealth of published data on this. WRc[3] quote the data in Table 12.1. This illustrates two of the key points relating to disinfection using chlorine:

- pH is important. This is because hypochlorous acid is much more effective than the hypochlorite ion. Thus the pH value for disinfection using chlorine should be no higher than 8.0. Above this hypochlorous acid increasingly dissociates into hydrogen and hypochlorite ions;
- free chlorine is much more effective than combined chlorine.

Table 12.1. Contact time required for 99% destruction of E. coli[3]

	Concentration (mg/l)	Contact time (min)	pH	Temperature (°C)
Free chlorine	0.1	0.4	6	5
	0.1	6	8.5	20–29
	0.4	1	8.5	20–29
Monochloramine	0.1	60	4.5	15
	1	6	4.5	15
Dichloramine	1	64	9	15

WHO[5] suggest that effective chlorination requires: a free chlorine residual of more than 0.5 mg/l; a contact time of at least 30 min; a turbidity of less than 1 NTU; and a pH of no more than 8.0. This is equivalent to a 'Ct' of 15.

US standards[6] require a three-log removal/inactivation of *Giardia* cysts and a four-log removal/reduction in viruses—both reductions being totals applied to filtration and disinfection. Where disinfection alone is used, without filtration or clarification, a Ct of 6 is required for a four-log inactivation of viruses at 10°C and a pH value of 6–8, and a Ct of 104 for a three-log inactivation of *Giardia*.

ULTRAVIOLET DISINFECTION

UV disinfection involves passing water through high-intensity UV radiation. The radiation kills or inactivates bacteria and viruses by affecting their RNA and DNA. UV radiation at a wavelength of around 260 nm is normally used. Water is passed through a chamber or channel containing UV lamps which expose the water to a controlled dose of UV radiation for a minimum period. The UV radiation penetrates organisms and initiates photochemical reactions within the cells inhibiting or killing the organisms. It is very effective against bacteria and viruses but less effective or ineffective against larger protozoa. The radiation is produced by lamps that produce light at a wavelength dependent on the type of lamp used.

Key parameters in UV disinfection are the power intensity (mW/cm^2) and the dose applied (mWs/cm^2), the intensity multiplied by the exposure time. A minimum dose is 15 mWs/cm^2, which would normally produce water of acceptable bacteriological quality, but 25–40 mWs/cm^2 is desirable.

For UV disinfection to be effective the water has to have high transmissivity for the UV used. The transmissivity or adsorbence of water is measured to assess the suitability of water for UV disinfection and to estimate if a higher dose is required to account for adsorbence within the water. Normally there are no problems with potable water but where iron levels are above 0.1 mg/l or water is hard,[7] there can be problems with scaling of the lamps. Ideally a water sample should be tested before using UV disinfection; the most likely problem with drinking water is scaling.

UV disinfection is particularly useful for groundwater sources which pump directly into supply and which do not have a chlorine contact tank. UV disinfection provides assurance that the water leaving the site is safe, with a residual disinfectant required only to maintain the safety of the water.

OZONATION

Ozonation has been covered in Chapter 11. It is a powerful disinfectant but is not normally used solely for disinfection in the UK. The dosages and contact times required for effective disinfection are lower than those needed for treatment of organic chemicals and thus where ozone is used for other purposes it also provides very effective disinfection. For disinfection an ozone residual of 0.4 mg/l and a contact time of 4 min will effectively kill all bacteria and viruses of concern, providing there is either plug flow or two cells to prevent short-circuiting.

OTHER METHODS OF DISINFECTION

The methods covered above are those normally used in conventional water treatment. Other processes that may be encountered include:

- bleaching powder—consists largely of calcium hypochlorite;
- boiling water—suitable for emergency use only;
- use of metal ions—the use of silver ions to disinfect water is possible, this is only suitable for emergency short-term use for potable water;
- potassium permanganate—can be used but is not very effective;
- membrane processes—ultrafiltration and reverse osmosis membranes remove micro-organisms;
- bromine or iodine—emergency use only.

Chlorine dosing equipment

Traditionally chlorine has been dosed using gaseous chlorine, in cylinders containing around 70 kg of chlorine used on smaller works or using drums containing up to 1000 kg on larger works. Where gaseous chlorine is dosed the system consists of the following:

- Chlorine store with chlorine gas detection equipment and automatic shutdown.
- Automatic changeover equipment on the drums/cylinders.

- Gas evaporator for plants using a large quantity of chlorine.
  If the rate at which the chlorine leaves the cylinder is too
  high, the rapid transition from liquid to gas causes extremely
  low temperature, which, in turn, causes chlorine crystals to
  form which block the feed pipe. One remedy is to lower the
  rate of flow leaving the cylinder by increasing the number
  of cylinders connected. The critical rate of draw-off varies
  with the size of the cylinder. A 1000 kg drum can discharge
  at up to 10 kg/h in average temperatures but a 70 kg
  cylinder will 'freeze' at very much lower rates. There is no
  theoretical limit to the number of cylinders that can be
  connected in parallel, but in practice, in big works, where
  flows exceed 60 kg/h it becomes cheaper to insert a
  special 'evaporator' between the cylinder and the
  regulating valve.
- A regulating valve to reduce the gas pressure down to
  around 150–200 millibar.
- A gas control valve to control the rate of gas supplied to
  the injection unit—which is known as an 'Ejector'.
- The gas ejector which injects the chlorine into a supply.
  The gas is injected into a supply of high-pressure motive
  water using an eductor, a venturi system which creates a
  vacuum which sucks in the chlorine.
- An injection point where the motive water with a high
  concentration of chlorine is injected into the water to be
  chlorinated.
- A control system comprising a flowmeter and downstream
  sampling and analysis equipment used by the residual
  controller to control the chlorine gas rate by the gas control
  valve. Different equipment suppliers use different control
  systems and the design of such systems should be done by the
  supplier.

Specialist suppliers supply chlorine systems. The concentrated
chlorine solution is corrosive, and all piping and pumps have to be of
suitably resistant materials. Figure 12.3 is a diagram of a typical
chlorination system. The gas piping is largely under vacuum, minimizing the risk of any chlorine gas escaping. There needs to be a
supply of high-pressure motive water, typically at a pressure of two
or three bars higher than the pressure at the point of injection into

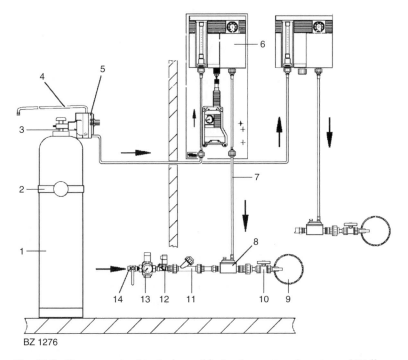

BZ 1276

Fig. 12.3. Components of typical gas chlorination system (courtesy of Wallace and Tiernan) (1) Chlorine gas cylinder; (2) cylinder retainer; (3) cylinder valve; (4) vent; (5) vacuum demand valve; (6) chlorine control unit; (7) chlorine suction pipe; (8) injector (vacuum eductor); (9) water main to be chlorinated; (10) main injection unit; (11) non-return valve; (12) solenoid valve; (13) pressure reducing valve; (14) isolation valve for carrier water

the main flow. Chlorine gas is injected into this motive water, by the gas injection system, and the motive water is then introduced into the water to be chlorinated. The motive water may be supplied by dedicated pumps or may use water from the treated-water delivery mains leaving the site. Whichever is used it is absolutely vital that the supply is continuous.

Where a treatment works uses super/dechlorination or chloramination there will be similar systems for either sulfur dioxide or ammonia.

Large gaseous chlorine installations are recognized as a major risk to life. A large escape of gas could kill people in the vicinity of the works and increasingly tight regulations apply to works using gaseous chlorine. In England works storing large quantities of chlorine come under the COMAH Regulations, which also apply to major chemical works, and similar regulations apply elsewhere in the UK and overseas. Concerns over the possible effects of leaks and the regulations themselves are leading to some water companies changing from chlorine gas to using either bulk sodium hypochlorite or an on-site electrolytic chlorination (OSEC) system.

Control and safety measures

Chlorinators are very reliable but as a check on the amount of chlorine being used it is common practice to mount the gas cylinders on a weighing machine. The daily loss of weight of liquid chlorine can be compared with the amount used as indicated by the chlorinator setting and any discrepancy investigated.

Chlorine is a deadly gas and extensive safety precautions are required. In the last few years there has been a complete change of thought about safety measures. It used to be considered that the chances of an accident were negligible and that the amount of gas to be dealt with would be small. It was common practice to give chlorine rooms adequate ventilation at low level, because chlorine is heavier than air, and all doors opened directly to the outside and not to internal rooms. However, now provision is always made for a more serious accident, envisaging a large-scale escape of gas, and it is the general consensus that it is advisable to:

- house chlorine apparatus in gas-tight rooms;
- install sensors that can detect chlorine in the atmosphere; and
- provide equipment—with extractor fans, chlorine-neutralizing chemicals and sprays which start automatically when the sensors signal that chlorine gas concentrations in excess of 2 ppm are present in the atmosphere, however, the extractor fans should be stopped in the event of a major leak, to prevent large quantities of chlorine being passed into the atmosphere.

Chlorine safety measures are best left to specialists.

SODIUM HYPOCHLORITE AND OSEC SYSTEMS

Sodium hypochlorite solution is available in bulk for large works, or in 45 l containers for small works. Where used it is dosed like any other liquid chemical using a standard chemical dosing pump with flow control and dosing controlled by the chlorine residual. Sodium hypochlorite comes in a standard strength solution with a nominal chlorine concentration of up to 15% available chlorine. However, the solution deteriorates over time and storage times should be limited to no more than 2 months' supply.

OSEC involves electrolysis of a solution of sodium chloride solution to produce sodium hypochlorite solution. Hydrogen is also produced as a waste product. The process produces sodium hypochlorite solution with a chlorine content of 6–9%. The system uses bulk dry sodium chloride, which is dissolved in a saturator using softened water. Approximately 3.5 kg of salt is required to produce 1 kg of chlorine. This represents approximately 50% of the theoretical potential production from the salt; the balance of the salt remains in the hypochlorite solution and is dosed into the treated water. Power use is around 5 kW h/kg of chlorine. OSEC is becoming widely used on large plants, but some companies prefer to use bulk sodium hypochlorite, which may be more expensive but is easier to operate.

Potential problems with OSEC include the production of chlorate, for which the UK Government sets a limit of 0.7 mg/l,[4] and bromate from bromide present as an impurity in the salt, for which the water-quality standard is 10 μg/l of bromate.

Monitoring and testing

In the UK, all treated water leaving a treatment works is continuously monitored for chlorine residual, normally using a triple validation instrument. There will be high and low chlorine concentration alarms, and normally a 'low low' alarm that will shut down the works if it is maintained for more than a few minutes. Monitors and alarms should be regularly checked and calibrated.

CONTACT AND TREATED-WATER TANKS

Where water has been chlorinated it passes to a contact tank. This should have a fixed volume and will typically be sized to provide

30–60 min retention at full plant capacity. The contact tank should have a more or less constant volume independent of flow, with an outlet weir to maintain water level under low-flow conditions or when the downstream hydraulic head is low. The tank should be baffled to prevent both short-circuiting and dead zones. A typical contact tank for a small or medium sized works would be approximately square with two internal baffle walls and a 60-min retention time. Water would enter in one corner and flow round the baffles to an outlet weir located in the far corner from the inlet. Such an arrangement would ensure an acceptable contact time after allowing for short-circuiting. For larger plants it becomes more economical to reduce the storage to nearer 30 min while optimizing the design to reduce short-circuiting.

After the water leaves the contact tank, the chlorine dose would be adjusted, normally trimming by dechlorination, before passing to the treated-water storage and thence to distribution.

13: Waterworks wastes and sludges

INTRODUCTION

All water-treatment processes produce wastes of some form and this chapter considers the liquid wastes produced by the main processes of clarification and filtration. These wastes comprise the bulk of the waste produced by water-treatment plants. Wastewater from some other processes which involve backwashing, notably GAC adsorption, will also normally be treated with these wastes. Other wastes are also briefly considered.

A 1998 survey on waterworks sludge production in the UK[1] gave the figures in Table 13.1 for sludge production and disposal. The survey is of interest in that it shows disposal to landfill was the predominant disposal route followed by disposal to a foul sewer. However, the works discharging to a sewer were all operated by companies or authorities responsible for both water and sewerage services, and apparently the water-treatment plants were not charged for discharging their wastes to the sewer. If the economic cost were to be charged, disposal to sewer would almost certainly reduce. This is because sewage treatment does not significantly reduce the quantity of solids that needs to be disposed of, but it does involve diluting the waterworks wastes with sewage, passing them through the sewage-treatment works, and then still having to dispose of the solids in the sewage sludge.

Another reason for treating wastes is that where lagoons are used the land occupied is large and often valuable or required for other purposes. Also, process improvements may mean a larger quantity of waste is produced leading to capacity problems in lagoons. There may also be difficulties in complying with the quality requirements of a new consent. Therefore for a variety of reasons it will become unusual in the UK to have anything other than proper treatment for the wastes produced, with recycling of water recovered from the wastes. However, in other areas of the world lagoons may well still be appropriate as a destination for sludge.

Table 13.1. Estimated WTW sludge production and disposal routes for the UK in 1998

Type of sludge	Production (dry solids)		Disposal (percentage of each type of sludge)				
	Quantity (t/year)	Percentage	Landfill	Foul sewer	Lagoon	Novel management	Other
Natural	5 850	4.5	1.6	59.6	5.9	15.8	17.0
Alum coagulation	58 528	44.8	51.1	34.9	0.9	4.8	8.2
Iron coagulation	43 040	32.9	56.9	20.4	4.2	13.5	5.1
Softening	22 240	17.0	89.5	0.0	0.0	7.8	2.7
Other	1 098	0.8	28.2	4.0	0.0	0.0	67.8
Total	130 756	100					

Notes:
1. Based on UKWIR report 'Sewage Sludge SL–09' 1999.[1]
2. Total water production estimated as 16 500 Mld.

In the UK, most liquid discharges from WTWs are classified as trade effluent and thus where the discharge is to controlled water (defined in Section 104 of the Water Resources Act 1991), consent is required from the Environment Agency (or the Scottish Environmental Protection Agency); where the discharge is to a sewer, a trade effluent, consent is required from the sewerage undertaker. Wastes that are disposed of to a landfill are classified as industrial waste and are controlled wastes; they therefore fall under the requirements of Part II of the Environmental Protection Act (EPA) 1990, and the Waste Management Licensing Regulations (WMLR) 1994. A waste management licence is therefore needed to dispose of the waste.

In the past it was quite normal for little attention to be paid to wastewaters. Often they were sent to lagoons or disposed of to a sewer and effectively forgotten. This is clearly no longer possible. Wastes can conveniently be thought of as falling into four main categories:

- liquid chemical wastes;
- washings from screens and microstrainers;
- dirty washwaters from filters and adsorbers;
- sludges.

These will be considered in turn after consideration of sludge quantities.

EXPRESSION OF SLUDGE QUANTITIES

The solids concentration of sludge should be expressed as a weight (w) of dry solids per unit volume (v), as w/v, or a proportion of the weight of sludge that is dry solids, as w/w. The solids concentrations found in liquid water-treatment sludges are from around 1 to 50 kg/m^3, with most at the lower end. At this level of solids the specific gravity of the sludge is approximately 1, and 1 l of sludge weighs approximately 1 kg. In practice it is usual to see solids contents expressed as a percentage. This is ambiguous as, for example, 1% DS could mean that 1% of the sludge is solids either by volume or by weight. In practice it is often used both as being the percentage of a weight of sludge that is solid material (i.e. w/w), and as the weight of sludge solid material per unit volume (i.e. w/v). The former is a correct usage whereas the latter assumes a specific gravity of 1 for the sludge and thus is incorrect. However, at low sludge solids concentrations the specific gravity of a sludge is approximately 1 and there is no significant error in expressing, for example, a sludge with 10 000 mg DS/l as 1% DS.

In accordance with common practice, percentages have been used in this chapter and unless otherwise indicated the percentage indicates w/v, with $x\%$ meaning x g of dry solids/100 ml. If the basis is weight of dry solids per weight of sludge this is indicated by the symbol w/w as in $x\%$ (w/w).

ESTIMATION OF SLUDGE QUANTITIES

The sludge produced by a water-treatment plant is derived from six sources:

- suspended solids in the raw water;
- colour that is removed during treatment;
- dissolved chemicals that precipitate during the process—mainly dissolved iron and manganese, and any hardness precipitated during softening;
- coagulants added during the process that precipitate out during treatment;
- other chemicals added during treatment such as PAC and bentonite;
- biological growth within the processes—difficult to estimate and not normally significant except for slow sand filters.

There are two methods of calculating sludge quantities. They are the calculation of dry solids production based on raw-water quality and likely chemical dosages during treatment, and applying rules of thumb to estimate sludge volumes as a proportion of water treated. The former is the proper way, but such are the uncertainties that it is always sensible to check the reasonableness of the result using the latter method.

Suspended solids

If there is data on raw-water suspended solids then this is easy. However, for water-treatment it is more common to have turbidity data. In the absence of any other data suspended solids (in mg/l) can be assumed to be twice the turbidity (in NTU).

Colour

Precipitated colour contributes to sludge quantities. Sludge production (in mg/l) of between 10% and 20% of the colour removed (degrees Hazen) are quoted; WRc have recommended 20%.[2]

Dissolved chemicals

Assuming that iron precipitates as ferric hydroxide, each mg/l of iron removed produces 1.9 mg/l of dry solids.

Where water is softened using lime and soda the solids production is given by:

$$\text{solids (mg/l)} = \text{lime dose} + 1.26 \times \text{decrease in hardness} \\ - 0.26 \times \text{sodium carbonate dose}$$

where dosages are in mg/l and hardness is mg $CaCO_3$/l.

Coagulant

Assuming that iron and metal coagulants precipitate as hydroxides the quantities of sludge produced by metal coagulants are as given in Table 13.2. Polymer doses are removed as dosed; however, these are low and are unlikely to be significant.

Other chemicals

Where powdered activated sludge or bentonite is added, the sludge solids will increase by the applied dose.

Thus if the quality of the raw water is known together with the likely coagulant dosage it is possible to estimate the quantity of dry solids that will be produced as sludge during treatment. However

Table 13.2. Solids production from different coagulants

Coagulant dose expressed as mg/l of	Solids production per mg/l of coagulant
Al	2.9 mg/l
Al_2O_3	1.5 mg/l
Fe	1.9 mg/l
Fe_2O_3	1.35 mg/l

the estimate will only be as good as the data on raw-water quality, and the accuracy of chemical dosages required during treatment. Quantities will vary with variations in raw-water quality and a judgement will need to be made as to the quantities used for the design of sludge handling and treatment facilities. Having calculated the theoretical quantities it is always useful to check the results against rules of thumb for sludge production.

RULES OF THUMB

It is common to use rules of thumb for estimating sludge quantities for the two main sources of sludge and wastewater, clarifiers and filters. Quantities are clearly not accurate but reflect the order of magnitude of production expected.

Clarifier sludge

Clarifiers produce sludges with solids contents typically between 0.1% and 0.8%. The lower values apply to raw waters with low colour and turbidity and the higher values to waters with high turbidity treated in clarifiers using sludge cones. A typical solids concentration from a modern clarifier is 0.2–0.3%. Volumes of clarifier sludge normally range from 1% to 2.5% of throughput; values above 3.5% indicate an unusual water or operational problems. WRc[2] quotes an average of 2.2% for aluminium sludge and 1.8% for ferric sludge. The higher values for alum may reflect overdosing of alum for pH control.

DAF produces sludge with a solids content of around 2% dry solids or higher; thus the quantity of sludge, though not of solids, is lower than for clarifiers.

There are two ways to calculate the weight of solids removed in clarifiers. Clarification should produce water with a suspended solids value of around 5 mg/l,[3] with the balance of the solids removed by the clarifiers. Alternatively, WRc[2] quotes removals of 50–90% of solids in clarifiers with a typical value of 70%.

Filter washwater

In a normal works all remaining solids are removed in the filters. Filter backwash water typically contains about 0.03% of solids and the volumes range from as low as 1% up to 5% of throughput, with a typical value of around 3%. It is common for the quantity to vary if raw-water and clarified-water quantity vary over time causing more frequent backwashing at times of poor water quality. It is increasingly common for there to be detailed backwashing requirements in specifications for rapid gravity filters and these often result in rates of backwash water production at the higher end of the numbers quoted above.

Liquid chemical wastes

Liquid chemical wastes are discharged from ion-exchange plants used for softening or nitrate removal which are regenerated with brine. Depending on the degree of treatment provided up to around 5% of the total throughput of a softening plant may be discharged as waste, with a much lower proportion for a nitrate-removal plant. The wastes will contain the ions removed by treatment, mainly calcium and magnesium from a softening plant or sulfate and nitrate from a nitrate-removal plant, together with sodium and chloride.

In the UK, such wastes generally have to be discharged to the public sewers as it is not possible to obtain a permit to discharge such wastes to surface waters.

Reverse osmosis and nanofiltration also produce a liquid chemical waste. This is difficult to dispose of and is commonly best discharged to a sewer. However, in hot arid countries evaporation lagoons may be another option.

Debris from screens and microstrainers

Screens produce small quantities of easily drained waste. This is easily disposed of. Microstrainers produce waste that will easily dewater to a consistency where it can be handled and disposed of. In practice, waste from both screens and microstrainers is not normally a problem.

DIRTY WASHWATERS FROM FILTERS AND ADSORBERS

Filters and adsorbers require regular backwashing. The washwaters from both are very dilute and normal practice is to take them to tanks which function both as flow-balancing tanks and settlement tanks. The solids are allowed to settle and the supernatant is recycled. A typical arrangement would be to have two tanks, each with the capacity for several filter backwashes. Once a tank was full backwash water would be diverted to the second tank, and the contents of the first tank allowed to settle. After a period of 2 or more hours of quiescent settlement supernatant would be removed, typically by a floating arm draw-off, and pumped back to the head of the works. Settled sludge would be pumped to the sludge-treatment works. The tank would then be available to receive more flow. For large works with a large number of filters it is more economic to have some form of upflow settlement tank that would operate intermittently, as filters were backwashed. Smaller works might only have a single tank capable of holding the water from a full set of filter backwashes.

Another option is not to settle but simply to use a flow-balancing tank that allows the washwater to be returned at a steady rate to the head of the works. This simpler option is now less common as it involves recycling all material removed in the filters, including *Cryptosporidium* oocysts. The maximum return rate is generally stated to be 5% of the flow being treated. If there is no provision to deal with settled solids, there will need to be some means of mixing the tank contents to prevent settlement.

The sludge from backwash settling tanks will typically have a solids concentration of around 1%.

Sludges

In a conventional water-treatment plant, the main source of sludge is the clarification stage. Some additional sludge may be generated from the settlement of filter backwash water. Assuming the sludge is not discharged to a sewer it will normally go to a lagoon or else be thickened and dewatered.

Sludge from softening is of a different nature to coagulation sludge, having a higher inorganic content of inorganic material. It is considered separately later.

Thickening

When sludge is dosed with polymer, at the correct rate in a suitable mixer which breaks up any existing flocs, and allowed to settle, flocs reform and the sludge separates into a clear supernatant and a layer of flocs with a clearly defined boundary between the two. If settling is allowed to continue the layer of flocs will consolidate such that all free-water escapes. A sludge thickener is normally designed to achieve this state of no free-water. The sludge from such a process will have a solids content that depends on the nature of the sludge and may be more than 10% (*w/w*) although 5–10% is more common.

On large works thickening will be by means of a continuous thickener. This is a circular tank, either flat-bottomed or with a gentle fall towards the centre, with a conical sludge hopper in the centre. The thickener contains a rotating sludge rake that draws sludge towards the central hopper. The tank is typically 2–3.5 m deep and with a diameter to give a hydraulic loading of 1.5 m/h. The depth depends on the nature of the sludge and how easily it thickens. Pilot trials are required to determine this, although providing the tank is at least 3 m deep it can normally be operated satisfactorily.

Sludge is dosed with polymer and introduced into a central feed well at a depth of around 1 m below water surface level. Flocs form as the sludge enters the thickener and on entry the flocs start to settle and the free-water to rise. The water rises up and is collected over an outlet weir. The flocs settle, and as they settle they are compressed and dewatered so that by the time they reach the bottom of the tank all free-water has been released. The thickened sludge is then withdrawn, normally to a holding tank. There is a distinct interface between the sludge and the supernatant, and the process is controlled by withdrawing sludge to maintain the interface at the required level. The supernatant is normally clear with low turbidity.

Providing the tank is properly sized and operated, the solids content of the thickened sludge is independent of the solids content of the feed to the thickener, depending solely on the nature of the sludge itself.

Two points to note are that polymer dosing is essential for the satisfactory operation of these thickeners and that in the UK, there are restrictions on polymer dosing where the supernatant is returned to the main process stream and enters the water supply.[4]

Dewatering and disposal

The dewatering process used should be considered in parallel with the ultimate disposal route. The cost of thickening increases with the solids content of the sludge produced and the designer should therefore consider the solids content required. This depends on the ultimate disposal route including transport distance. In practice, most sludge goes to a landfill and the requirement is a solids content of more than 20% in order that the sludge can be handled as a solid waste. The performance of the various forms of mechanical dewatering plant with different sludges varies widely and is susceptible to process changes which alter the nature of the sludge. Performance claims therefore need to be treated with some scepticism, and should be supported by site testing, ideally in a pilot plant.

Filter plate presses

Plate presses comprise a horizontal stack of steel or heavy-duty plastic plates, normally square, each with a recess on either side. In each recess there is a mat of filter cloth so that in the gap between two adjacent plates there are two mats of filter cloth. The plates are mounted in a frame which allows them to be separated when required. The principle of the filter press is that the plates are locked together. Sludge is then pumped into the gap between each set of filter cloths. The cloth retains solids and the liquid passes though the cloth and is drained away through openings in the plates. As the filter mat becomes coated with solids it becomes more efficient at retaining the solids and the pressure in the system rises. Sludge is pumped into the press over a period of several hours, after which the recesses are full of dewatered sludge, known as cake. The solids content of the cake depends on the nature of the sludge, the pressure used, and to a limited extent on the time of pressing. Typically a solids content of around 25% (*w/w*) can be achieved. In general, filter plate presses produce cake with the highest-solids content compared to other commonly used thickening processes.

At the end of the charging cycle the press is allowed to stand for a while to further dewater. The pressure is then released and the plates separated allowing the cake to be knocked off the filter cloths. This drops down into a skip or on to a conveyor. The plates are then brought back together and the process is repeated. It is normal to have two or three cycles per day.

Older installations require significant operator input, particularly for removing the sludge from the cloths, but fully automatic presses are common now. As this is a batch process balancing storage is required.

Centrifuges

Centrifuges can be used to dewater waterworks sludges. They consist of a horizontal cylinder that spins at a speed of around 4000 rpm, giving an acceleration of around 2000g. Sludge is introduced into the centrifuge at one end and is thrown out onto the inside of the revolving bowl. The centrifuge is designed such that there is a preset thickness of sludge and liquid on the bowl. Sludge solids that collect on the inside of the bowl are moved towards the cake discharge point by a screw conveyor that has a small clearance to the spinning bowl. The liquid from the sludge flows the other way over a circular outlet weir fixed to one end of the bowl. As the sludge moves along the bowl the acceleration forces cause the solids to be compressed and lose water.

The key process parameters are the spinning speed; the rates of feeding and withdrawing sludge; and polyelectrolyte dose.

Filter belt presses

These consist of two looped belts made of filter cloth that are mounted on a series of rollers. Sludge is introduced into the space between the belts and is then carried through the press. This is designed such that the rollers progressively squeeze the sludge allowing the liquid to pass through the cloth and leaving the cake on the cloth from where it is scraped off. The process is controlled by the rate of sludge application, the speed of the belts, and the polyelectrolyte dose.

Vacuum filtration

Vacuum filters may be of either the drum or horizontal type. They operate by applying a vacuum to one side of a filter mat; the vacuum then withdraws water through the filter mat from the sludge being filtered.

The drum type comprises a horizontal cylindrical drum covered in filter cloth supported on a structural frame. The ends of the cylinder are sealed but have a liquid drain and a suction pipe passing through. The lower part of the drum sits in a tank of sludge. The drum is rotated while a vacuum is applied to the inside. The drum picks up sludge from the tank and as it rotates liquid is drawn out of the sludge. Just before the drum enters the sludge, after a complete revolution, sludge is scraped off the filter cloth.

The performance of drum vacuum filters depends on the speed of rotation; the depth of immersion, which affects the thickness of sludge on the drum, and the conditioning of the sludge. Clearly there is a limitation on the vacuum that can be applied. These filters often perform poorly with gelatinous coagulation sludge but may perform well with sludge from softening. Even when used with softening sludge their performance may deteriorate if polymer is dosed.

A variation on the drum vacuum filter is the horizontal vacuum filter. These use a looped horizontal filter cloth, with sludge introduced onto one end of the horizontal belt.

Lagoons and drying beds

For lagoons to work satisfactorily a capacity equal to around 200 days of waste sludge production is required. If the quantity of waste produced was 2% of plant throughput then the volume would be four times the daily flow treated. Lagoons are typically 1 m deep. The solids content of the sludge taken out from lagoons is rarely significantly above 10%.

Drying beds are constructed of a level bed of sand resting on an underdrainage system of gravel and drainage pipes. The beds have shallow division walls and a system of pipes or channels to convey the sludge to the beds. It is normal to thicken the sludge before passing it to the drying beds. After drying the sludge is removed and taken for disposal. In the UK, an area of around $60 \, m^2/1000 \, m^3/day$ of plant capacity is appropriate. In hot dry climates this can be significantly reduced and drying beds are often very satisfactory.

Softening sludge

The lime–soda process produces a mineral sludge, except where surface waters are softened when there may be a significant organic content. The mineral sludges are relatively easy to dewater providing no polymer has been used. The sludge is, however, very thixotropic, and even at high-solids contents may liquefy if disturbed. This makes it more difficult to handle and transport. Softening sludge containing polymer has a more gelatinous nature and this can make it more difficult to dewater. The ideal sludge from an operator's point of view is that produced from pellet reactors. This consists of hard free-draining pellets.

Disposal

This is normally to a landfill. However, there is increasing interest in using the sludge in ways that involve re-use. Such uses include:

- incorporation into construction materials such as brick, cement, or manufactured aggregate;
- incorporation into soils or soil improvers;
- use in land reclamation or agriculture.

To date there is no clear economically viable option to landfill. This is partly because of regulations covering the classification of waterworks sludge. However in the future, increased landfill costs and any relaxation in the regulations covering water-treatment sludge may make one or more of the above more economic.

14: Water demand and use

INTRODUCTION

Nowadays, in countries with limited water resources and increasing private sector involvement in water supply, it is often both unacceptable and uneconomical to design a water-treatment plant to treat a greater quantity of water than will be required within a reasonable number of years. More than this, it is often economic to manage demand for water to a level that minimizes the investment in new water-resources and treatment plant. This chapter presents an introduction to water demand and losses from water systems. Demand is considered in very broad terms with the aim of identifying the major components and giving an indication of their magnitude, and how demand varies over a year. Unaccounted-for-water is discussed in broad terms and the issues relating to it are covered in outline.

WATER DEMAND

Water demand and use is a subject that has become very complex over the past few years. This reflects its increasing importance and the amount of effort spent in studying it.

In the UK, until 1988 water-supply systems were generally publicly owned, domestic water supplies in the UK were unmetered, and it was widely accepted that increased incomes and improved standards of living automatically meant increased water use. Demand was unmanaged and increased at a steady rate. It was assumed that countries with low water usage would in the fullness of time reach North American levels of prosperity and water usage. Water-supply systems were designed with generous amounts of spare capacity.

Things are very different now. This has come about as a result of public resistance to new reservoirs, of increasing concern over low flows in rivers and streams as a consequence of perceived over-abstraction, and the unwillingness of the water-industry regulator to sanction increases in the price of water to fund the cost of new water-resources works. Now water suppliers have duties to manage demand and to continuously monitor water resources and demands. In the UK, it is difficult to find new usable water resources in many parts of the country. The 1995 Environment Act[1] places a legal requirement on English and Welsh water companies to manage demand through the promotion of efficient water use. In many areas, there is pressure to reduce abstractions, most notably where groundwater abstraction is perceived either to have led to formerly perennial rivers to become seasonal, or to have led to occasionally unacceptably low flows in perennial rivers. Even if water resources were available, the suppliers are now commercial companies that are unwilling to invest in facilities unless they will produce an acceptable financial return on the investment. Regulatory limitations on water prices may mean that suppliers will be unable to recoup all or part of the costs associated with developing new water resources. This new restrictive attitude to water resources has meant a sea change in attitudes towards demand and use of water. The role of the water supplier is becoming one of not simply meeting demand, but partly of managing demand within the available water resources. The development and use of new resources is increasingly the last resort, rather than the first.

Partly, as a result of the change in attitudes towards meeting demand, there has also been a change in attitudes towards paying for water. In the UK, it is now increasingly accepted that water should normally be paid for on the basis of metered use. This is supported by the EU Water Framework Directive which requires that 'Member States shall ensure by 2010 that water pricing policies provide adequate incentives for users to use water resources efficiently and thereby contribute to the environmental policies of this Directive'. All new houses in England and Wales now have meters. Most non-domestic supplies are now metered. Under the Water Industry Act 1999,[2] water companies can compulsorily meter customers with high non-essential use of water, such as those with a swimming pool or garden sprinkler. As well as compulsory metering of potential high users of water, there is also a voluntary move towards metering. Most householders in England and Wales still pay for water by a formula based on the rateable value of their property. Occupants of houses of higher rateable value who pay a high flat rate but use a relatively small quantity of water would therefore be better-off with a metered supply. Such people are increasingly opting for a metered supply. Once a house is metered, there will be much greater incentive to conserve water and, when buying new appliances, to choose those with less water usage.

In the UK, the proportion of metered domestic properties is expected to rise from around 17% in 2000 to around 30% in 2005. One result of increased metering is that there is increasingly accurate information on water consumption, and on water losses. This information enables demands to be predicted more accurately.

Components of demand

In practical terms, the demand of a water system is the quantity of water that has to be put into the system to meet the rate at which water is withdrawn. There are five main components to this demand:

- losses in the trunk main and distribution system;
- losses in the customers' supply pipes;
- domestic use—including garden and other external uses;
- industrial and commercial use, including offices, hospitals, and government; and
- operational use by the water supplier.

The latter three components are largely independent of the pressure within the water-distribution system, but the first two are closely pressure related.

LOSSES FROM WATER SYSTEMS

All water-distribution systems 'lose' water. The most obvious loss of water is through leaks. However, from the supplier's point of view losses cover all water for which no payment is received, such as operational use and water taken illegally.

Leakage is water that is lost from the distribution system through leaks. It is simple to define, but surprisingly difficult to measure and compare. In England and Wales, water companies have to report total leakage including supply pipe losses on the customers' properties. Even where the supply to a customer is through a metered connection, the water that leaks after the meter but prior to the supply entering the property is considered as leakage. English and Welsh water companies have to report detailed statistics and total leakage each year in Table 10 of their annual return to OFWAT, the water-industry regulator. In this table, the leakage is calculated from 16 inputs, a measure of the complexity of the subject.

Leakage is a function of five main factors (Lambert *et al.*[3]):

- the length and condition of the distribution mains;
- the number of customer connections;
- system pressure;
- the location of customer meters;
- the number of hours of water supply.

Leakage is actually a rate of loss of water. However, giving an absolute value to leakage is of no use whatsoever as a comparator of a performance indicator, except for a particular system. To know that a particular town is losing $100\,m^3/h$ in leakage offers very little illumination on the performance of the town's distribution system. The headline way of expressing leakage is to express it as a percentage of the water put into the distribution system. This is straightforward and readily understood but again not very useful. It takes little account of the five factors listed above that affect leakage. For example, consider two cities one of which has a distribution system in good condition and which provides a continuous supply of water at high-distribution pressure; and the other which supplies water through a system in poor condition for only a few hours a day at very low pressure. The first city may well lose a similar percentage of the water entering supply as the second city, indicating similar performance. However, consumers in the two cities would see the performance of the systems very differently.

Also, a city system will have a high number of consumers in a small area served by a relatively short length of distribution main. On the other hand, a rural system will have a large length of main per customer connection. If the condition of the mains and the house connections are the same for both the city and the rural case, and the systems operate at the same pressure, the rural system will inevitably have a much higher absolute level of leakage, and thus a worse rate when expressed as a percentage of water supplied.

Thus current thinking[3] on the best international practice for measuring leakage is to introduce three key concepts:

- Leakage is best expressed as litres/service connection/day, when the system is pressurized. This, however, takes no account of system pressure, length of main, hours of supply, or location of meters.

- There is a minimum unavoidable leakage rate for a system, the base level of leakage. This accepts that leakage is always with us, even in the best-constructed and -maintained systems. The minimum unavoidable leakage can be calculated for a particular system based on mains length, system pressure, number of connections, hours of supply, and position of water meters. The calculation has to be done using appropriate factors for the various leakage components. The unavoidable leakage can also be expressed in litres/service connection/day, when the system is pressurized.
- By dividing actual leakage rates with the minimum unavoidable leakage rate, the relative performance of a system can be defined. A value of 'one' indicates a system operating at the minimum attainable leakage rate, whereas a value of 'two' indicates it would be technically possible to halve leakage. Well-managed systems have a ratio of between 1.2 and 1.8.[3]

This of course is a fine theoretical approach to leakage. In reality, it is expensive to maintain leakage at low levels. In practice, each system will have an economic level of leakage, at which the costs of maintaining leakage at that particular level are balanced by the costs of supplying the additional water that will leak from the system. Thus, a supplier that has a low-cost (and abundant) supply of water would normally have a higher economic level of leakage than a supplier that has a high-cost (or limited) supply of water.

Water suppliers often have more practical interests other than leakage alone. A water supplier is often more interested in 'non-revenue' water. Clearly, to maximize revenue, a water supplier wants as much water as possible to be 'revenue' water, water for which the supplier receives payment. This is a much more complex concept than leakage. Revenue water is the quantity of water that is measured at the point of delivery to a customer, plus non-metered water legitimately used by a customer. 'Non-revenue' water includes leakage in the distribution system, leakage in unmetered customers' supply pipes, meter-under registration, and water taken from the system and used without payment—which could be illegal use, operational use such as mains flushing, or fire-fighting. A key problem in defining both leakage and non-revenue water is estimating consumption by unmetered properties. This requires suppliers to estimate the unmetered per capita consumption (pcc) by means of a monitoring study of a discrete area. Total water use can be monitored within a defined area and the population can be estimated from census data. From this data, together with knowledge of metered consumption and minimum night flows, the leakage and unmetered pcc will be estimated.

Measurement of leakage

Leakage measurement developed greatly over the 1990s. At the beginning of the period, leakage was normally measured by isolating a water-supply zone supplied by a service reservoir at night. Water entering the zone was then measured as accurately as possible over a period of a few hours in the early hours of the morning, often by measuring the rate of fall of water level in a water tower or service reservoir. This gave the total volume supplied over the period of the test. The volume used by large metered users would then be deducted from this total, based on meter readings before and after the test. The quantity of water assumed to be used for legitimate domestic purposes would then be deducted, leaving a quantity of water assumed to be leakage. The most obvious problem with this method is that the figure assumed for legitimate night use was often the most important and critical item. A small change in this figure would greatly affect the leakage quantity. The normal figure used for legitimate night use was originally 6 l/property/h, based on research carried out in the 1980s. However, this was reduced to a suggested figure of 1.7 l/property/h in the 1996 'Managing Leakage' document[4] and currently a figure of around 2 l/property/h is widely used in the UK water industry. This reduction in the allowance for legitimate night use from 6 to 2 l/property/h would have greatly increased the reported leakage rate in any zone and is one of the factors that means that it is not normally possible to make comparisons between current and reported historic leakage rates.

Nowadays, in the UK it is usual to use accurate on-line meters and automatic recording and comparison of night flows using telemetry systems to monitor actual instantaneous flow rates. Large water-supply zones have been divided into smaller district meter areas

(DMAs). This has been done partly to avoid large pressure variations in zones, and partly to allow leakage rates to be more accurately monitored. A DMA will have a limited number of feeds, each of which will have on-line flow measurement. This allows monitoring of minimum night flows and permits the supplier to detect any sudden increase, perhaps due to a burst pipe, and any longer-term increase, indicative of an increasing number of smaller leaks. DMAs typically serve between 1000 and 10000 properties. This size range has been derived from the need to establish the smallest number of zones, to minimize costs, while maintaining the ability to detect pipe bursts. In a large zone, smaller pipe bursts will not be quickly detected.

PEAK FLOWS

Demands are normally quoted as specified absolute figures. However, domestic demands vary throughout the year, mainly dependent on temperature and rainfall. In the summer, there is more water use in houses for bathing, which is dependent on temperature. There is also more external use of water in the summer, mainly for garden watering, and this is very sensitive to rainfall. The nature of an area affects the size of the seasonal variation. Seaside resorts have higher peak to average demands than city areas. However, there may also be very high flows in the winter, normally during extended cold spells. These flows are associated with a high rate of winter pipe bursts, particularly in areas with clay soils.

Table 14.1 gives some typical values of some of the numbers used in demand forecasting in England and Wales. The numbers are not absolute but are indicative only. Of particular note are the typical ratios of peak flows to average flows. It is usual to size the water-treatment works to supply the average day peak week demand. This is insufficient to meet the peak daily demand, and the difference between the two is provided by water stored in service reservoirs. Thus treatment works are sized based on the average daily flow in the peak week, service reservoirs on peak day demands, and distribution systems on maximum instantaneous flows.

DEMAND MANAGEMENT

In the UK, there is increasing consideration of demand management. This has been driven largely by the 1995 Environment Act,

Table 14.1. Typical values of key parameters used in demand forecasting

Parameter	Typical value
Domestic use	150 l/head/day (metered supply)
Hospital use	300–500 l/patient/day
Office use	70 l/staff/day
Hotel use	400–500 l/guest/day
Average day peak week demand	120–140% of annual average demand
Peak daily demand	140–160% of annual average demand
Leakage (typical range for well-managed large system)	15–25% of water put into supply
Leakage (typical range for well-managed system)	0.8–1.2 l/connection/m of system pressure/day (derived from Lambert *et al.*[3])

and partly by pragmatic need. Given that in practice there are limited water resources available, how does one manage demand such that it can be met? Demand management requires a comprehensive knowledge of all issues relating to water use. Factors include cost, the consumption of water-using domestic appliances, the condition of the distribution system, social changes, and climate and weather.

Suppliers are therefore adopting a variety of techniques to manage (limit) demand. Techniques being used or currently being considered include:

- Tariff—tariffs can be designed to manage demand. For instance, it is possible to have rising block tariffs such that those who use little water have a low unit charge rate and those who use much pay much more for the extra water. It is also possible to have seasonal tariffs such that water used during times when the water resources are not stressed, normally during the winter, are charged less; whereas water used when the water resources are stressed, during the summer, are charged at a much higher rate. Variable tariffs will become more practical as meter penetration increases and remote meter reading is introduced.
- Metering of households with non-domestic use—the use of garden hoses and swimming pools are non-domestic uses. The metering of consumers who are considered discretionary

high users is required in England and Wales by the 1999
Water Industries Act.

- Publicity and education—aimed at achieving greater
awareness of the value and importance of water and to
encourage lower use and less wastage.
- Selective restrictions—these include such measures as
hosepipe and sprinkler bans but are only normally used
during droughts.
- Regulation of water using appliances—the 1999 Water Bylaws
of England and Wales[5] regulate the usage of water of certain
appliances such as WCs, washing machines, and dishwashers.
As an instance, the size of the WC cistern has been reduced
from 9 to 6 l. The new Regulations for washing machines
require a use per cycle not exceeding 35 l, whereas most
current machines have a usage of 50–90 l. For dishwashers,
the figures are 54 l/cycle, compared to most current machines
which use 20–30 l.
- Lower use fittings—as well as more efficient appliances,
significant reductions in water usage could be made by
installing spray head taps instead of conventional ones,
showers instead of baths (but power showers can use as much
as a bath) and sensors on urinals and other appliances which
would then provide water only when required.
- encouragement of recycling by industrial users—it is
increasingly economic for industrial users to pay for water
audits, which identify wastage, processes that could be made
more water efficient, and where recycling can be utilized.
- Domestic water audits—water audits are a good way of
identifying both wastage and where more efficient use could be
made of water. These are best carried out by trained
personnel, but this is not cost-effective. The cost-effective
way is to send a pro-forma to users, for them to
complete—but this results in much lower benefits.
- Leakage reduction—English and Welsh water suppliers now
have to assess and achieve their economic level of leakage.
This is monitored by the water-industry regulator. Total
leakage in 1994/5 was 5112 Ml/day but by 1999/2000 this had
reduced to 3306 Ml/day.[6] This improvement was a result of
great efforts by the water companies. The methods used were
reducing the pressure in the water-distribution network

wherever possible; establishing DMAs to monitor flows, using data loggers to identify new leaks; using leak noise correlators to identify the location of leaks; and the prompt repair of leaks and pipe bursts.

As well as the above, future possibilities include:

- Water re-use in households—about one-third of household use is for flushing the WC. This function does not need high-quality water. Experiments are being made to recycle grey water from baths, showers, and wash-hand basins. This needs some treatment to remove impurities and some disinfection to ensure no scum build up in the pan. It also needs some storage and generally a pump. This system is currently too expensive to retro-fit into existing properties. For new properties, it still needs 'operation' and is not yet 'fit and forget'. As such few installations have been made. A similar alternative is the use of rainwater from the roof.
- Re-use of effluent—Article 12 of the Urban Wastewater Directive indicates that treated wastewater should be re-used wherever appropriate. Currently only Windhoek in Namibia directly returns treated sewage effluent to the water-treatment works.

The 1995 Environment Act requires water suppliers in England and Wales to produce Water Efficiency Plans to promote the efficient use of water by its customers and these strategies are reviewed annually by the industry's commercial regulator, OFWAT.

Research has shown that demand management can significantly affect water consumption. Installation of metering has been shown to reduce average water demand by around 10% and peak demand by up to 30%. This scale of reduction is significant in its own right, either reducing the capacities of any new water-supply infrastructure needed or, more commonly, postponing the need for demand driven new investment. Figure 14.1, based on data from the Environment Agency, shows the average daily amount of potable water put into the distribution system since 1960, by the English and Welsh water companies. The quantity therefore includes water lost through leakage. It can be seen that the annual increase in water demand has levelled out and started to fall. It remains to be seen what happens in the future but there is little doubt that use will not increase as it has in the past.

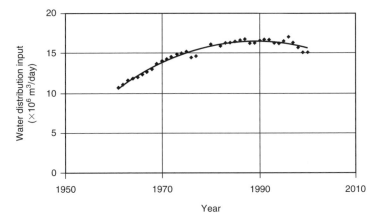

Fig. 14.1. Water demand in England and Wales 1961–2000

WATER-RESOURCE PLANNING

Demands and demand management obviously need to be considered within the restraints of a particular supplier's water resources. Water companies in England and Wales are required to produce an annual assessment of available water resources and projected demands over a 25-year period for review and agreement with the industry regulator.

These assessments establish the long-term need for water-resource development within a particular supply area. The supplier's plans take into account actual and projected changes in demand and available resources, including the probable effects of climate change.

The demand forecasts for households can be prepared by projecting the levels of ownership of water-using appliances, the frequency of their use, and the water consumption per use (or cycle for washing machines). Projected demand management savings such as those achieved from the installation of new household meters and leakage reductions to economic levels are also allowed for. Where new resources are found to be required, following the implementation of demand management measures, the supplier's long-term resource development profile is established by prioritizing new schemes, such as new boreholes and storage reservoirs, in least cost order.

Appendix 1: Sample calculations

1. EXAMPLES OF CALCULATIONS FOR COAGULATION

Example 1. Sinuous channel

A sinuous channel has 15 round-the-end cross-walls. Water is passed along with a velocity of 0.2 m/s between the cross-walls and 0.5 m/s round the ends. The flow is 0.3 m³/s and the nominal retention time is 25 min. For a temperature of 10°C, estimate the additional loss of head, the power dissipated, and the G and GT values (Fig. A1.1).

$$\text{Additional head loss} \atop \text{(given by } v^2/2g) = \frac{16 \times 0.2^2 + 15 \times 0.5^2 \text{ m}}{2 \times 9.81} = 0.224 \text{ m}$$

Using equation (5.2):

$$P = 0.3 \times 9.80 \times 10^3 \times 0.224 \text{ W} = 659 \text{ W}$$

and

$$T = 25 \times 60 \text{ s} = 1500 \text{ s}$$
$$V = 0.3 \times 1500 \text{ m}^3 = 450 \text{ m}^3.$$

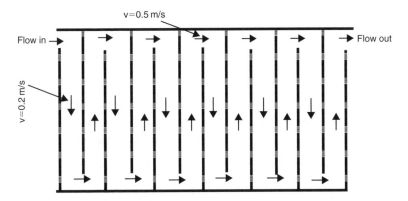

Fig. A1.1. Sinuous channel for flocculation

Using equation (5.1):

$$G = [659/(1.31 \times 10^{-3} \times 450)]^{0.5} \, s^{-1} = 33.4 \, s^{-1}.$$

Thus

$$GT = 33.4 \times 1500 = 5.01 \times 10^4.$$

Example 2. Weir mixing

A flow of $50\,000 \, m^3$/day is to be dosed with a coagulant. Destabilization will be by double layer compression. To ensure proper mixing a G of $1500 \, s^{-1}$ is required. Assuming that the flow over the weir discharges into a chamber with a volume of $2 \, m^3$, what height of weir is required?

$$\text{Flow} = 0.58 \, m^3/s.$$

Therefore

retention time in mixing chamber $= 2 \, m^3/(0.58 \, m^3/s) = 3.4 \, s.$

Apply equation (5.3):

$$G = (\rho gh/\mu T)^{0.5}.$$

Thus

$$h = G^2 \times \mu \times T/\rho g$$
$$h = 1500^2 \times 1.31 \times 10^{-3} \times 3.4/(9.8 \times 10^3).$$

Thus

$$h = 1.02 \, m.$$

2. EXAMPLES OF CALCULATION OF CHEMICAL DOSAGES

A water-treatment works treats a flow of $50\,000 \, m^3$/day. The water is dosed with alum at a dose of 5 mg Al/l. This permits the removal of an average of 5 mg/l of colloidal material in the clarifiers. Calculate:

1. the quantity of 8% Al_2O_3 alum solution required (specific gravity of 8% solution is 1.32);
2. the amount of alkalinity required to react with the alum; and

3. the volume of sludge produced assuming a sludge solids content of 2.0% (on a weight basis) and a specific gravity of 2.2 for the sludge solids.

1. The quantity of aluminium dosed per day is:

$$50\,000\,m^3 \times 5\,mg\,Al/l \times 10^3\,l/m^3 \times kg/10^6\,mg = 250\,kg\,Al/day.$$

The molecular weight of Al_2O_3 is 102 and the atomic weight of aluminium is 27; therefore 54 kg Al is equivalent to 102 kg Al_2O_3.

Thus 250 kg Al is equivalent to

$$250 \times 102/54\,kg\,Al_2O_3/day = 472.2\,kg/day$$

The alum is being dosed as an 8% Al_2O_3 solution. Thus the weight of alum solution being dosed is:

$$472.2\,kg\,Al_2O_3/day \times 1/0.08 = 5903\,kg/day.$$

At a specific gravity of 1.32 this is equivalent to a solution volume of: $5903/(1.3 \times 1000)\,m^3/day = 4.47\,m^3/day$.

2. The reaction between alum and natural alkalinity is given by equation (6.1). Adding molecular weights we get:

$$Al_2(SO_4)_3 \cdot 14H_2O + 3Ca(HCO_3)_2 \rightleftharpoons 2Al(OH)_3 + 3CaSO_4 + 6CO_2 + 14H_2O$$

$$594 \qquad\qquad 3 \times 162 \qquad 2 \times 78 \quad 3 \times 136 \quad 6 \times 44 \quad 14 \times 18$$

Thus 2 mol Al (with a molecular weight of 2×27) are equivalent to 1 mol $Al_2(SO_4)_3$ and require 3 mol $Ca(HCO_3)_2$ to react. Thus on a weight basis:

1 g Al will react with $(3 \times 162)/(2 \times 27) = 7.56$ g $Ca(HCO_3)_2$.

Thus 5 mg/l Al will react with 37.8 mg/l $Ca(HCO_3)_2$. Expressing this as $CaCO_3$ it is equivalent to:

37.8 × equivalent weight $CaCO_3$/equivalent weight $Ca(HCO_3)_2$
= 37.8 × 50/81 = 23.3 mg/l as $CaCO_3$.

3. The quantity of sludge produced by the solids removed from the raw water is given by:

$$5\,mg/l \times 50\,000\,m^3/day \times 10^3\,l/m^3 \times kg/10^6\,mg = 250\,kg/day.$$

In addition the quantity of hydroxide floc can be calculated from equation (6.1).

1 mol Al (with a molecular weight of 27) produces 1 mol Al(OH)$_3$ with a molecular weight of 78. Thus on a weight basis:

$$1 \text{ g Al will produce } 78/27 \text{ g Al (OH)}_3 = 2.9 \text{ g}.$$

Thus 250 kg will produce 725 kg Al(OH)$_3$. Thus total weight of solids produced is 975 kg/day. At 2% solids this is equivalent to a volume of sludge of:

$$975/(1000 \times 2.2) + 975 \times (98/2000) = 48.21 \text{ m}^3/\text{day}.$$

Note that for sludges having a low solids content an acceptable answer is obtained by simply assuming that the quantity can be calculated on the basis of a specific gravity of 1 for the sludge solids. In this case the volume calculated on this basis would be $975 \times 100/2 \times 1/1000 = 48.75 \text{ m}^3/\text{day}$. The difference is small, representing the difference in volume between 975 kg of solids at a specific gravity of 2.2 compared to a specific gravity of 1.

3. EXAMPLE OF CALCULATION OF DETAIL DESIGN OF PLATE SETTLER—SIMPLIFIED EXAMPLE

As an example of the factors to be taken into account, the following example is a simplified procedure to design plate clarifiers to treat a flow 'Q' of 250 000 m^3/day (10 416 m^3/h), a very large flow equivalent to a population of the order of 1 million. Assume a particle settling velocity of 0.6 m/h (a low value for alum floc). For a surface-loading rate (Q/A) of 0.6 m/h the effective area 'A' of plate required is:

$$A = 10\,416/0.6 \text{ m}^2 = 17\,361 \text{ m}^2.$$

Assume that the plates are inclined at 60°, and are 2 m long and 5 m wide. Then the projected area of one plate, 'a', would be:

$$a = 5 \times 2 \times \cos 60° = 5 \text{ m}^2.$$

Thus a total of 3472 plates would be required.

The clarifiers require quiescent settling and as stated in Chapter 8, this requires a Reynolds number of less than 800.

$$Re = \rho V L / \eta$$

where
    Re = Reynolds number
    V = velocity of the water

η = dynamic viscosity (temperature dependent—assume minimum water temperature of 0°C = 0.001791)

L = the spacing between the plates.

Consider a plate spacing of 100 mm. Then

$$V = 10\,416/3600/(0.1 \times 5)/3472\,\text{m/s} = 0.001667\,\text{m/s}$$
$$Re = 1000 \times 0.001667 \times 0.1/0.001791 = 93$$

which is satisfactory.

In practice, the plate area would be increased to allow for entry and exit effects, typically by around 25% to 4340 plates.

It is of interest to calculate the net surface-loading rate on an installation like this. Clearly this depends on such factors as plate thickness and the number of tanks used. Assume that 16 tanks are provided and plate thickness is 10 mm. The surface area of each tank, which would contain 272 plates, would be:

$$5 \times (272 \times (0.1 + 0.01)/\cos 30° + 2 \times \cos 60°) = 178\,\text{m}^2.$$

Thus total area is 2848 m². Thus the apparent surface loading is 250 000/2848/24 m/h, which is 3.65 m³/m²/h. This is more than six times the loading that would be possible for the floc considered in a conventional rectangular settlement tank.

The retention time in the settlers would be approximately 20, or 25 min if the area of plate were to be increased by 25%. Note that although decreasing plate spacing would not affect the Reynolds number, it would increase flow velocity, and decrease retention time, requiring better process control in order to react more quickly to water quality changes or process upsets.

4. EXAMPLES OF CALCULATION OF CHEMICAL DOSAGES REQUIRED FOR SODA ASH SOFTENING

In order to calculate the quantities of lime and sodium carbonate required to soften a given water it is necessary to know the concentrations of magnesium and calcium, the concentrations of the anions, including bicarbonate, and the pH of the water. This allows the calculation of calcium and magnesium carbonate and non-carbonate hardness. Consider, for example, water with the following analysis:

Parameter	Concentration (mg/l)
Sodium	20
Potassium	30
Calcium	5
Magnesium	10
Chloride	40
Bicarbonate HCO_3^-	67
Sulfate	5
Nitrate	10

The first step is to construct an equivalence table.

Species	Molecular weight (g)	Equivalent weight (g)	Concentration		Cumulative concentration (g equivs/l)
			mg/l	g equivs/l	
Ca^{2+}	40	20	32	0.00160	0.00160
Mg^{2+}	24.3	12.15	15	0.00123	0.00283
Na^+	23	23	12	0.00052	0.00336
K^+	29.1	29.1	10	0.00034	0.00370
HCO_3^{2-}	61	30.5	65	0.00213	0.00213
SO_4^{2-}	96	48	5	0.00010	0.00224
NO_3^{2-}	62	31	10	0.00032	0.00256
Cl^-	35.5	35.5	40	0.00113	0.00368

The final column shows the anions and cations to be approximately in balance, as would be expected, or at least hoped. By constructing an equivalence diagram (Fig. A1.2) it can be seen that the hardness is as follows:

Type of hardness	Concentration (g equivs/l)
Calcium carbonate hardness	0.00160
Magnesium carbonate hardness	0.00053
Calcium non-carbonate hardness	0
Magnesium non-carbonate hardness	0.00070

Finally the quantities of lime and sodium carbonate can be calculated from the equations set out above and the equivalent weights of lime (74/2 = 37) and sodium carbonate (106/2 = 53).

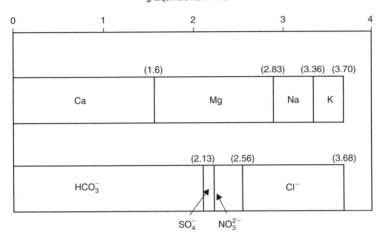

Fig. A1.2. Equivalence diagram

Type of hardness	Concentration (g equivs/l)	Lime required		Sodium carbonate required	
		g equivs/l	mg/l	g equivs/l	mg/l
Calcium carbonate	0.00160	0.00160	59.2	0	0
Magnesium carbonate	0.00053	0.00106	39.2	0	0
Calcium non-carbonate	0	0	0	0	0
Magnesium non-carbonate	0.00070	0.00070	25.9	0.00070	37.1
Total		0.00336	124.3	0.00070	37.1

The above represents a simple calculation. It ignores the effect of dissolved CO_2 which will create a lime demand in addition to that calculated above. The free CO_2 can be found by means of a fairly complex calculation if the pH value, temperature and TDS are known.

Appendix 2: Comparison of different drinking water-quality standards

The following table summarizes, and simplifies, the requirements of the drinking water-quality standards listed to allow simple comparisons to be made. However, the requirements and recommendations within the standards quoted are complex and cannot be fully summarized in a simple table. In order to fully understand the requirements of the standards quoted, it is essential to refer to the source document. The listing is incomplete, particularly with respect to American Standards for organic chemicals.

The European Directive and British legislation includes mandatory parameters and indicator parameters.

USA Standards are classified into three categories: Primary health-based standards that are mandatory for all states; secondary standards, affecting aesthetics or that cause cosmetic effects, that are not mandatory; and health advisories, estimates of acceptable concentrations that are not legally enforceable. The system is complex and changes as standards are introduced and revised on the basis of additional knowledge, particularly for organic chemicals. In the table, only primary maximum contaminant levels (MCLs) and Secondary Drinking Water Regulation values are listed. The primary parameters also have guideline MCLs. For some parameters, the standard is a treatment technique rather than a numeric value. However, the technique is also associated with a maximum allowable concentration for the parameter.

Parameter	England and Wales Water Supply (Water Quality) Regulations 2000	European Drinking Water Directive (98/83/EC of 3 November 1998)	WHO Guidelines for drinking water quality	USA Drinking Water Standards (EPA 822-B-00-001 Summer 2000)
Microbiological Parameters				
Enterococci (No./100 ml)	0	0	—	—
Escherichia coli (No./100 ml)	0	0	0	0
Coliform bacteria (No./100 ml)	0 (Note 2)	0	0	0 (in 95% of samples)
Clostridium perfringens (No./100 ml)	0	0	—	—
Colony counts (No./100 ml)	No abnormal change	No abnormal change	—	TT (500/ml) (Note 9)
Cryptosporidium parvum (Average No./10 l)	1	—	—	TT (99% removal)
Physical parameters				
Colour (Mg/l Pt/Co)	20	Acceptable with no abnormal change	15 TCU	15 (Note 10)
Turbidity (NTU)	4/1 (Note 1)	Acceptable with no abnormal change	5	1 (95° < 0.3)
Taste (Dilution)	3 at 25°C	Acceptable with no abnormal change	Acceptable	—
Odour (Dilution)	3 at 25°C	Acceptable with no abnormal change	Acceptable	3 (Note 10)
pH	6.5–10.0	6.5–9.5	<8.0 for disinfection	6.5–8.5 (Note 10)
Conductivity (μS/cm)	2500	2500	—	—
Dissolved solids (mg/l)	—	—	1000	500 (Note 10)

Chemical parameters			
Aluminium (µg/l)	200	200	50–200 (Note 10)
Antimony (µg/l)	5	5	6
Arsenic (µg/l)	10	10	5
Boron (µg/l)	1	1	—
Cadmium (µg/l)	5	5	5
Chromium (µg/l)	50	50	100
Copper (µg/l)	2000	2000	TT (action level of 1300)
Iron (µg/l)	200	300	300 (Note 10)
Lead (µg/l)	25/10 (Note 3)	10	TT (action level of 15)
Manganese (µg/l)	50	500	50 (Note 10)
Mercury (µg/l)	1	1	2
Nickel (µg/l)	20	20	—
Selenium (µg/l)	10	10	50
Sodium (mg/l)	200	200	—
Ammonium (mgNH$_4$/l)	0.5	1.5	—
Bromate (µg/l)	10	10	10
Chloride (mg/l)	250	250	250 (Note 10)
Cyanide (µg/l)	50	70	200
Fluoride (mg/l)	1.5	1.5	4
Nitrate (mg/l)	50 (Note 8)	50	10 mg/l as nitrogen
Nitrite (mg/l)	0.5	3	1 mg/l as nitrogen
Sulfate (mg/l)	250	250	250 (Note 10)
Acrylamide (µg/l)	0.10	—	TT
Benzene (µg/l)	1.0	—	5
Benzo(*a*)pyrene (µg/l)	0.010	0.70	0.2
1,2 dichloroethane (µg/l)	3	30	5

Parameter	England and Wales Water Supply (Water Quality) Regulations 2000	European Drinking Water Directive (98/83/EC of 3 November 1998)	WHO Guidelines for drinking water quality	USA Drinking Water Standards (EPA 822-B-00-001 Summer 2000)
Epichlorohydrin (µg/l)	0.10	0.10	0.40	TT
Polycyclic aromatic hydrocarbons (Note 4) µg/l	0.10	0.10	0.10	0.2 (Note 11)
Tetrachloroethene (µg/l)	10 (Note 5)	10 (Note 5)	40	—
Tetrachloromethane (µg/l)	3	—		—
Trichloroethene (µg/l)	10 (Note 5)	10 (Note 5)	70	—
Trihalomethanes (total) (Note 6) (µg/l)	100	100	—	80
Vinyl chloride (µg/l)	0.50	0.50	—	2
Pesticides (total) (µg/l)	0.50 (Note 7)	0.50 (Note 7)	—	—
Aldrin (µg/l)	0.03	—	0.03	—
Dieldrin (µg/l)	0.03	—	—	—
Heptachlor (µg/l)	0.03	—	0.03	0.4
Heptachlor epoxide (µg/l)	0.03	—	0.03	0.2
Pesticides (other) (µg/l)	0.10 (Note 7)	0.10 (Note 7)	—	—
Tritium (Bq/l)	100	100	—	—

Notes:

1. Value of 1 applies at treatment works; value of 4 at consumers taps.
2. Applies at water-treatment works and service reservoirs. At service reservoirs at least 95% of samples must be zero.
3. 25 µg/l from 25/12/03 to 25/12/13, 10 µg/l from 25/12/13.
4. Covers benzo(*b*)fluoranthene; benzo(*k*)fluoranthene; benzo(*ghi*)perylene; and indeno(1,2,3-cd)pyrene.
5. The standard of 10 µg/l applies to the total concentration of tetrachloroethane and trichloroethane.

6. Comprises chloroform, bromoform, dibromochloromethane, and bromodichloromethane.
7. The total is the sum of the concentrations of all pesticides. The 'other' standard applies to any pesticide save for the four named pesticides.
8. Additionally [concentration of nitrate in mg/l]/50 + [concentration of nitrite in mg/l]/3 must be less than or equal to one.
9. TT denotes treatment technique.
10. Secondary standard—not legally enforceable at Federal level.
11. Relates to benzo(a)pyrene.

Appendix 3: Glossary

Absorption, Adsorption   Similar words often loosely used interchangeably. Adsorption is a surface effect in which the adsorbate (the substance being removed by the process of adsorption) is adsorbed onto the surface of the adsorbent. The process is normally reversible, allowing the adsorbent to be removed and regenerated. Specific surface area and porosity are critically important in the performance of many adsorption processes. Absorption is a process where the substance being removed is taken within the absorbent; it is not a surface effect.

Acidity   A term used loosely to denote how acidic water is. Water with a pH value of less than 7 is termed acidic, meaning that it has a higher concentration of hydrogen ions than neutral water, which has a pH value of 7. Acidity may be due to natural causes such as the presence of free CO_2 or humic or other naturally occurring acids, but it can also result from industrial pollution. Any water with a pH value below 7 is termed acidic. Waters with low pH values are normally corrosive.

Activated silica   A coagulant aid which is prepared from sodium silicate 'activated' by various chemicals which include chlorine, sodium bicarbonate, sulfuric and hydrochloric acid, and CO_2. These act as neutralizing agents on the sodium silicate to form a silica sol which may improve the performance of the primary coagulant. It is little used in potable water nowadays.

Activated carbon   A material consisting mainly of carbon. It is very porous with a very high specific area per unit of mass. It adsorbs many organic chemicals and is used for treatment of tastes and odours and pesticides. It can be manufactured from coal, peat, coconut shells or other materials with a high carbon content. Activated carbon is used either as powder (PAC), when it is dosed prior to clarification, or in granular form (GAC), when it is used in beds as an adsorbent. GAC can also be used as the granular media in rapid gravity filters or as a layer in slow sand filters.

Algae   These are primitive, autotrophic, plant-like, single cell micro-organisms. They use sunlight and photosynthesis to grow; typically they contain cellulose within their cell walls. Because of this, their growth rates are relatively low. When conditions are right, single species grow in large numbers forming an algal bloom, with the numbers limited by the availability of nutrients, or by chemicals used within the algal cells. In drinking water treatment, they are often difficult to remove, blocking filters, and they are often associated with taste and odour problems.

Alkalinity   It has two meanings; it is either the equivalent of acidity for water with a pH value of more than 7, or more precisely as a measure of the total quantity of the dissolved substances in water that react with acids. For this latter definition, the alkalinity is usually expressed as $mgCaCO_3/l$ and is commonly measured by adding acid to water containing methyl orange indicator, which determines when the acidity is neutralized. In this case, it is sometimes written as alkalinity (M). The bicarbonates, carbonates, and hydroxides of calcium, magnesium, potassium, and sodium cause alkalinity. High alkalinity is often associated with high TDS and high pH. When the alkalinity equals the temporary hardness, no permanent hardness is present. If the alkalinity exceeds the total hardness, sodium bicarbonate must be present.

Alum   The commonly used word for aluminium sulfate.

Anion   A negatively charged ion.

Autotrophic   Refers to micro-organisms that do not need to use organic matter as a source of energy. The most common autotrophic organisms are those that use light for photosynthesis. See hetero-trophic.

Blue–green algae or cyanobacteria   A form of algae with some similarities to bacteria. Some blue–green algae produce toxins that are of concern.

Brownian motion Investigated by Robert Brown, an English botanist of the early nineteenth century, who observed that microscopic particles in suspension in a gas or a liquid had a continuous, random motion. It is one of the reasons why liquids in the same container tend to mix even if not stirred.

Cation A positively charged ion.

Clarifier Often used synonymously for a settlement basin. Some would argue that 'clarifier' is better used with reference to the flotation process, or solids contact basins that 'clarify' rather than 'settle'.

Clark's process A method of softening water. It involves adding lime, which reacts with the free carbonic acid and the carbonic acid combined in the bicarbonates of calcium and magnesium, to form insoluble carbonates and hydroxides.

Coagulant A substance which encourages coagulation.

Coagulation The process of destabilization of stable suspensions of fine material or colloidal particles in water. Coagulation followed by flocculation results in the formation of larger aggregated particles which are easier to remove during water treatment.

Coagulant aid A substance that assists coagulation, resulting in the production of larger, denser, or more stable flocs. Filter aids perform a similar function prior to filtration.

Co-current or counter-current Refers to the methods used to operate and regenerate adsorbers. Co-current systems operate such that the direction of normal operation (i.e. either vertically upwards or downwards) is the same as that in which regenerant is applied. For counter-current systems, regeneration is carried out in the opposite direction to normal flow. Counter-current systems are more efficient but in practice this is of little significance in potable-water treatment. It is, however, important in the production of ultra-pure water.

Coefficient of fineness The ratio between the dry silt by weight in milligrams per litre and the turbidity in NTUs, for any given sample of water. This gives an indication of the ease (or otherwise) of settling the solids in that water. For silt-laden water with a coefficient of fineness above unity, the solids will settle more quickly than for water with a coefficient below unity, because the higher figure indicates the presence of a higher concentration of fine sand.

Colloidal solutions (more correctly colloidal systems)   They are stable two-phase systems with fine particles of one phase dispersed in a second phase. Colloidal systems are normally thought of as solid particles suspended in a liquid (sols) but colloidal systems include liquid/liquid systems (emulsions) and gas/liquid systems (aerosol). The colloidal particles normally carry similar electric charges which prevent the particles from coagulating. In water treatment, colloidal systems are stable dilute suspensions of very small solid particles in water.

Colour   Water colour is caused by preferential adsorption of light of certain frequencies by water. True colour is the colour of water with no suspended material present. Apparent colour is/may be in part due to turbidity/suspended material. Can be expressed as mg/l Pt/Co, °Hazen, or True Colour Units (TCU), all of which are interchanmgeable for all practical purposes.

Complex   An ion formed by an ion combining with an atom or molecule. Particularly noteworthy in coagulation, where iron and aluminium sulfates form a series of short-lived complexes which can be very effective coagulants, before precipitating as hydroxides.

Crypto, *Cryptosporidium*   The everyday terms used for *Cryptosporidium parvum*, a protozoan parasite found in cattle and sheep which causes severe diarrhoea in humans. In healthy people, the diarrhoea lasts around a week. In immunosuppressed people, infants, and old and frail people, the infection can prove fatal. The organism is difficult to analyse for and it was only recognized as a problem relatively recently. Some humans act as carriers of the organism.

Direct filtration   The treatment of good-quality raw water by rapid gravity or pressure filtration without prior settlement or clarification. The water is normally coagulated and flocculated prior to filtration. In the UK, the process is generally considered unacceptable for surface waters because it is considered unsuitable for waters that may contain *Cryptosporidium*.

Dissolved air flotation (DAF) The process that introduces fine bubbles of air into water. The bubbles are derived from water super-saturated with air. The bubbles tend to form using solid particles as nuclei. The bubbles and the solids then float to the surface, where they are removed as a foam.

Enhanced coagulation A term derived from the USA surface water-treatment rules, which refer to coagulation enhanced by optimizing coagulation pH and ensuring rapid dispersion of the coagulant within the water being treated.

Eutrophication The process whereby nutrients and biological growth increase with time in a body of water. If nutrient input is unchecked, a lake may eventually fill with organic matter and become a bog. Eutrophication is stimulated by nutrients derived from fertilizer use in agriculture and from sewage effluents.

Epilimnion A result of thermal stratification of lakes and reservoirs during the summer. The epilimnion is the warm fully mixed upper layer of water. See thermal stratification.

Filtration A straining process for removing particles from water. Rapid gravity filters and pressure filters remove particles by predominantly physical means, whereas slow sand filters depend largely on biological action.

Flash mixer A device which mixes coagulant into the raw water with considerable violence induced either hydraulically or mechanically. The aim is to disperse the coagulant rapidly throughout the water being treated.

Floc The particles that are formed by the process of flocculation. These appear to the naked eye as wool-like, hence the name.

Flocculation The process which encourages coagulated (destabilized) particles to coalesce into larger particles. It involves stirring water in which floc has formed to induce the particles to coalesce and grow. The process can be optimized to produce the optimum size of particles required for efficient removal in subsequent treatment. For settlement, large particles are required; but for DAF or filtration, smaller flocs may be needed.

FTU   Formazin turbidity units—see Turbidity units.

G value   The stirring of water creates differences of velocity and therefore velocity gradients. The average temporal mean velocity gradient in a shearing fluid, denoted by 'G', is a factor used in calculating the energy input and size of mixing and flocculating chambers. GT is 'G' multiplied by the retention time, measured in seconds, in a mixing or flocculating chamber.

Giardia   *Giardia lamblia* is a protozoan organism that causes severe diarrhoea. It is carried by animals. Historically, it has been a problem in the USA, where it is associated with the use of water from protected catchments, untreated save for disinfection, for public water supply.

Hardness   A measure of the concentration of dissolved bivalent metals in water. In practical terms, these react with soap (but not with detergents) to form a scum. Until the hardness is neutralized, water will not form a lather with soap. Hardness also forms scale in hot-water systems. It is normally measured as $mgCaCO_3/l$. In France and Germany, hardness is also commonly expressed in degrees.
   $1\,mgCaCO_3/l = 0.1$ French Degree $= 0.056$ German Degree
   (The German degree is based on calcium oxide.)

Hazen units   A means of expressing the degree of colour in water. They are numerically the same as the platinum–cobalt scale.

Heterotrophic   Refers to organisms that use organic matter as a source of energy. Bacteria are heterotrophic organisms.

Hypolimnion   A result of thermal stratification of lakes and reservoirs during the summer. The hypolimnion is the cold unmixed lower layer of cold poor-quality water. See thermal stratification.

JTU   Jackson turbidity unit—see Turbidity units.

Lime–soda process   A softening process which uses lime and soda ash to precipitate calcium and magnesium, forming insoluble salts which can be settled and filtered from the water.

Langelier saturation index (LSI)   The difference between the actual pH of a water and the pH at which the water would be saturated with calcium carbonate (pH_s). The pH_s is calculated from the calcium concentration, temperature, TDS, and alkalinity of a water sample. If the LSI is positive, the water is at a pH greater than the pH_s, and calcium carbonate will precipitate from the water. This is commonly taken to denote that the water is non-aggressive, as it will coat pipes with a protective film of calcium carbonate. If the LSI is negative, the water will dissolve calcium carbonate, meaning that pipes will lose any protective layer of calcium carbonate. Such waters are regarded as corrosive. This is a simplistic approach to corrosion but nevertheless, in the absence of a simple theoretical measure of corrosiveness, it is commonly used as a treatment parameter.

Multi-layer filters   They are normally rapid gravity filters in which layers of different materials with different densities are used so that a coarser (but lighter) layer may remain on the surface. A commonly used combination is anthracite and sand. Although common sense suggests that the term could also apply to slow sand filters containing a layer of GAC, this is not common usage.

NTU   Nephelometric turbidity units—see Turbidity units.

Overflow rate   This relates the amount of water passing through a horizontal-flow sedimentation basin to its surface area. If 'Q' is flow in cubic metres per day and 'A' is area in square metres, then the overflow rate is 'Q/A' m/day.

Platinum–cobalt scale   A scale used for measuring colour in water against different concentrations of a platinum–cobalt solution. One unit of colour corresponds to one milligram of platinum per litre.

pH   A measure of the concentration of hydrogen ions in water, more correctly written as pH^+, where 'p' represents the concentration, and H^+ represents hydrogen ions. It is expressed as the negative logarithm to the base 10 of the concentration. Thus a pH of 7 represents a hydrogen ion concentration of 1×10^{-7} moles/l.

Pressure filter   Very similar in design and function to a rapid gravity sand filter, except that it is contained in a steel pressure vessel and can operate under pressure, if hydraulic conditions in the system require it to do so.

Rapid gravity filter A filter using granular filtration media that operates at relatively high-loading rates and which is cleaned by backwashing with air and water. Rapid gravity filters are the most commonly encountered form of filter used in potable-water treatment.

Schmutzdecke A German word meaning 'layer of dirt'. It is the biologically active layer which forms on the surface of a slow sand filter. It is a complex biological system normally containing both autotrophic and heterotrophic micro-organisms as well as more complex biological organisms. It also contains impurities removed by the filter from the water.

Slow sand filter This is the oldest form of sand filter, still fairly widely used. The water is passed very slowly downwards through sand beds. A biologically active layer, the schmutzdecke, forms on top of the sand. Treatment is largely biological. Cleaning involves removing the top layer of sand from the filter.

Thermal stratification During the summer deeper lakes and reservoirs tend to from two layers: an epilimnion, which is the upper warm fully mixed layer of water, and a hypolimnion, which is the cooler lower part of the lake or reservoir. The two layers are separated by a thermocline, across which there is a marked change in water temperature. The hypolimnion tends to be anoxic and to be high in dissolved iron and manganese. As winter approaches, the stratification breaks down and the lake or reservoir becomes fully mixed, often with a marked deterioration in water quality.

Tastes and odours They are closely related and arise in water for various reasons. Most waters have a slight taste which regular consumers do not notice. They are both measured as threshold odour numbers (TONs). The number represents the number of times a sample has to be diluted with pure water before the taste or odour cannot be detected. This is clearly a partly subjective measure.

Trihalomethanes (THMs) Compounds formed by the reaction between the halogen elements of chlorine and bromine and methane.

The methane (CH_4) molecule has three of its hydrogen atoms replaced by either bromine or chlorine. The compounds are chloroform, bromoform, dibromochloromethane, and bromodichloromethane.

Turbidity An optical effect caused by dispersion of and interference with light rays. It is caused by suspended solids but cannot be directly related to the quantity of solids present as it is also affected by the size, colour, and shape of the solids present. It is not to be confused with true colour which is independent of suspended solids.

Turbidity units Turbidity is expressed in the units applicable to the method of measurement or the instrument used (for example, JTUs, FTUs, NTUs). For all practical purposes, the units are numerically the same.

Uniformity coefficient A term often used to define the grading of filter sands. It is the ratio between the aperture openings passing 60% and 10% by weight of the sand sample.

Upward-flow filters They are rapid gravity filters in which the flow and the backwashing process proceed in an upward direction. They have the advantage of being able to store a large quantity of sediment in the coarser lower layers and thus require less backwashing. They are now very uncommon.

Appendix 4: SI units and basic conversion factors

SI BASE UNITS

Quantity/dimension	SI unit and symbol
Length	metre (m)
Mass	kilogram (kg)
Time	second (s)
Temperature	kelvin (K)
Electric current	ampere (A)
Amount of substance	moles (mol) (need also to specify associated atoms, molecules, or whatever)
Light intensity	candela (cd)

SOME SI-DERIVED UNITS WITH SPECIAL NAMES

Quantity/dimension	SI unit	Symbol	Dimensions
Force	newton	N	$kg\,m/s^2$
Pressure	pascal	Pa	N/m^2
Energy, work	joule	J	$N\,m$
Power	watt	W	J/s

ADDITIONAL UNITS RELEVANT TO WATER TREATMENT

Unit	Symbol
Area	m^2
Concentration (of amount of substance)	mol/m^3
Density	kg/m^3
Surface tension	N/m

Unit	Symbol
Velocity	m/s
Viscosity (dynamic)	Pa s
Viscosity (kinematic)	m^2/s
Volume	m^3

CONVERSION FACTORS

Mass	$1\,kg = 2.2\,lb$
Length	$1\,m = 3.208\,ft$
	$25.4\,mm = 1\,in$
Volume	$1\,m^3 = 35.3\,ft^3$
Liquid measure	$1\,m^3 = 220\,imp.\,gal = 263.7\,US\,gal$
Force	$1\,N = 0.225\,lb\,f$
Pressure	$1\,N/mm^2 = 145\,lb\,f/in^2$
	$1\,Pa = 1\,N/m^2 = 0.000145\,lb\,f/in^2$
Power	$1\,kW = 1.341\,HP$
Volume flows	$1\,m^3/s = 220.4\,imp.\,gal/s = 19.04\,Mgal/day$
Energy	$1\,J = 0.732\,ft\,lb\,f$
	$1\,kJ = 0.948\,Btu$

References

Chapter 1

1. BINNIE A. R. *Bradford Corporation Waterworks: prevention of waste of water.* Report dated 1 October 1885, Bradford.

Chapter 2

1. FOX C. S. *The geology of water supply.* Technical Press Ltd, London, 1949.
2. CHORUS I. and BARTRAM J. (Eds) *Toxic cyanobacteria in water.* E. & F.N. Spon on behalf of WHO, London, 1999, ISBN 0-419-23930-8.
3. COUNCIL OF EUROPEAN COMMUNITIES DIRECTIVE. 98/83/EC on the quality if water intended for human consumption. *EC Official J.* **L330**, 5/12/98.
4. DEGREMONT. *Water treatment handbook.* 1991.
5. WORLD HEALTH ORGANIZATION. *Guidelines for drinking water quality,* Vol. 1 *Recommendations.* WHO, Geneva, 1993 (and Addendum to Vol. 1 *Recommendations.* WHO, Geneva 1998).
6. *England and Wales: The Water Supply (Water Quality) Regulations 2000.* SI 2000, No. 3184, Stationery Office, London.
7. UK DRINKING WATER INSPECTORATE. *List of approved products and processes,* December 2000, www.dwi.detr.gov.uk/soslist.
8. USEPA. *National Primary and Secondary Drinking Water Regulations,* September 2000, www.epa.gov/safewater/mcl.htm/.
9. GRAHAM C. *Comparison of coliform/Escherichia coli count methods comparison trial conducted by the PHLS Water and Environmental Research Unit (Nottingham).* Statistics Report dated 21 April 1999.
10. WATER INDUSTRY. *England and Wales Water Supply (Water Quality) (Amendment) Regulations 1999.* SI 1524, HMSO, London.
11. FAWELL J. K. and MILLER D. G. Drinking water quality and the consumer. *JCIWEM* 1992, **6** (Dec.), 726–731.

Chapter 3

1. NICKOLS D. Personal communication, December 2000.
2. CAIRNCROSS S. and FEACHEM R. *Environmental health engineering in the tropics.* John Wiley and Sons, 1993, ISBN 0-471-93885-8.

3. WORLD HEALTH ORGANIZATION. *Guidelines for drinking water quality,* Vol. 1 *Recommendations.* WHO, Geneva, 1993.

Chapter 4

1. POYNTER S. F. B. and STEVENS J. K. The effects of storage on the bacteria of hygienic significance, in: *The effects of storage on water quality.* WRc, 1975.
2. OSKAM G. Main principles of water-quality improvement in reservoirs. *J. Water SRT—Aqua,* 1995, **44**(Suppl. 1), 23–29.
3. CROWE P. J. The Farmoor source works in operation. *J. Instn Wat. Engrs,* 1974, **28**(2), 123–124.
4. SAXTON K. J. H. Operation of the Grafham water scheme. *J. Instn Wat. Engrs,* 1970, **24**(7), 413.
5. SIMMONS, J. Algal control and destratification at Hanningfield Reservoir. *Water Sci. Technol.,* 1998, **37**, 2.
6. HALL T. and HYDE R. A. (Eds) *Water treatment processes and practices.* WRc, 1992.
7. COX C. R. *Operation and control of water treatment processes.* World Health Organization, Geneva, 1964, 212.
8. BULL A. W. and DARBY G. M. Sedimentation studies of turbid American rivers. *JAWWA,* 1928, **19**, 284–305.

Chapter 5

1. SAWYER C. N. and MCCARTY P. L. *Chemistry for environmental engineering.* McGraw Hill, 1978, ISBN 0-07-Y-66543-5.
2. AMIRTHARAJAH A. and MILLS K. M. Rapid mix design for mechanisms of alum coagulation. *JAWWA,* May 1982, 210–217.
3. CAMP T. R. and STEIN P. C. Velocity gradients and internal work in fluid motion. *J. Boston Soc. Engrs,* 1943, **XXX**(4), 219–237.
4. MEYER H. *Static mixers cut chemical costs World Water.* June 1995.
5. GHOSH M. M., COX C. D. and PRAKESH T. M. Polyelectrolyte selection for water treatment. *JAWWA,* March 1985, 67–73.
6. HUDSON H. *Water clarification processes.* Van Nostrand Reinhold, New York, 1981.

Chapter 6

1. BENFIELD L. D., JUDKINS J. F. and WEAND B. L. *Process chemistry for water and wastewater treatment.* Prentice Hall, 1982, ISBN 0-13-722975-5.
2. UK DRINKING WATER INSPECTORATE. *List of approved products and processes.* December 2000, www.dwi.detr.gov.uk/soslist.

Chapter 7

1. WEBBER N. B. *Fluid mechanics for civil engineers*. Chapman & Hall, 1971, ISBN 0-412-10600-0.
2. CAMP T. R. Sedimentation and design of settling tanks. *Trans. Am. Soc. Civ. Engrs*, 1946, **111**, 895–958.
3. NOVOTNY V., OLTHOF M., IMHOFF K. and, KRENKEL P. A. *Handbook of urban drainage and wastewater disposal*. John Wiley, 1989, ISBN 0-471-81037-1.
4. AWWA. *Water treatment plant design*. 1969.
5. MCLAUGHLIN R. T. The settling properties of suspensions. *JASCE*, 1959, **85**, 9–41.
6. YAO K. M. Theoretical study of high- rate sedimentation. *J. Water Pollut. Control Fedn*, 1970, **42**, 218.
7. YAO K. M. Design of high rate settlers. *JASCE, Env. Eng. Div*. 1973, **99**, 621–637.
8. EDZWALD J. K. *et al*. Flocculation and air requirements for dissolved air flotation. *JAWWWA*, 1992, **84**(3), 92–100.

Chapter 8

1. KAWAMURA S. Optimisation of basic water-treatment processes—design and operation: sedimentation and filtration. *J. Water SRT—Aqua*, 1996, **45**(3), 130–142.
2. KAWAMURA S. Hydraulic scale model simulation of the sedimentation process. *JAWWA* 1981, **73**(July), 372–379.
3. COX C. R. *Operation and control of water treatment processes*. World Health Organization, Geneva, 1964.
4. CULP G. L., WESNER and CULP R. L. *Handbook of public water supplies*. Van Nostrand, New York, 1986.
5. ZANONI A. E. and BLOMQUIST M. V. Column settling tests for flocculant suspensions. *Proc. ASCE*, June 1975, **101** EE3, 309–318.

Chapter 9

1. JAMES M. MONTGOMERY CONSULTING ENGINEERS INC. *Water treatment principles and design*. John Wiley & Sons, 1985, ISBN 0-471-4384-2.
2. BADENOCH J. *Cryptosporidium in water supplies. Report of the group of experts*. Department of the Environment, Department of Health, London, UK, HMSO, 1990.

3. BADENOCH J. *Cryptosporidium in water supplies. Second report of the group of experts.* Department of the Environment, Department of Health, London, UK, HMSO, 1995.
4. BOUCHIER I. *Cryptosporidium in water supplies. Third report of the group of experts.* Department of the Environment, Department of Health, London, UK, HMSO, 19
5. *The Water Supply (Water Quality) (Amendment) Regulations 1999.* SI No. 1524, HMSO, London.
6. KAWAMURA S. Optimisation of basic water-treatment processes—design and operation: sedimentation and filtration. *J. Water SRT—Aqua*, 1996, **45**(3), 130–142.

Chapter 10

1. CHERYAN M. *Ultrafiltration handbook.* Technomic Publishing Co, Lancaster Pennsylvania, 1986.
2. JACALANGELO J. G. and TRUSSELL R. R. Role of membrane technology in drinking water treatment in the United States. *Desalination*, 1997, **113**, 119–127.
3. IRVINE E., GROSE A. B. F., WELCH D. and Donn A. Nanofiltration for colour removal—7 years operational experience in Scotland. *Membrane Technology in Water and Wastewater Treatment.* Hillis P. (Ed.) Royal Society of Chemists, 2000, ISBN 0-85404-800-6.

Chapter 11

1. BENFIELD L. D., JUDKINS J. F., and Weand B. L. *Process chemistry for water and wastewater treatment.* Prentice Hall, 1982, ISBN 0-13-722975-5.
2. USEPA. *Alternative disinfectants and oxidants guidance manual.* EPA 815-R-99-014, April 1999.
3. MORRIS T. and SIVITER C. L. Application of a biological iron removal treatment process at Grove Water Treatment Works. *JCIWEM*, 2001, **15**(2), 117–121.
4. UNITED NATIONS SYNTHESIS. *Report on arsenic in drinking water*, 2001, www.who.int/water_sanitation_health/Arsenic/ArsenicUNReptoc.htm.
5. CARDEW P. T. Experience in the plumbosolvency control of soft waters and expectations for complying with the new lead standards. *Proc. Tech. Sem. Lead Drink. Water, CIWEM*, December 2000.

Chapter 12

1. RODMAN D. J., VAN DE VEER A. J. and HOLMES J. F. The pre-design of Berenplast Water Treatment Works, Rotterdam: Additional

processes to achieve biologically stable water. *JCIWEM* 1995, **9**, 344–352.

2. FAIR G. M., GEYER J. C. and OKUN D. A. *Elements of water supply and wastewater disposal.* John Wiley & Sons.
3. HALL T. and HYDE R. A. (Eds) *Water treatment processes and practice.* WRc, 1992.
4. UK DRINKING WATER INSPECTORATE. *List of approved products and processes*, December 2000, www.dwi.detr.gov.uk/soslist.
5. WORLD HEALTH ORGANIZATION. *Guidelines for drinking water quality*, Vol. 1 *Recommendations.* WHO, Geneva, 1993 (and Addendum to Vol. 1 *Recommendations.* WHO, Geneva, 1998).
6. *USA Surface Water Treatment Rule (SWTR)* (54 FR 27486, June 29 1989).
7. USEPA. *Alternative disinfectants and oxidants guidance manual.* EPA 815-R-99-014, April 1999.
8. WHITE C. G. *Handbook of Chlorination.* Van Nostrand Reinhold, New York, 1992.

Chapter 13

1. REPORT 'SEWAGE SLUDGE SL-09'. *Recycling of water treatment works sludges.* UKWIR London, 1999, ISBN 1-84057-167-5.
2. WARDEN J. H. Sludge treatment plant for waterworks. *Technical Report TR* 189, WRc, March 1983.
3. TWORT A. T., RATNAYAKA D. D. and BRANDT M. J. *Water supply*, 5th edn, Arnold, London, 2000.
4. UK DRINKING WATER INSPECTORATE. *List of approved products and processes*, December 2000, www.dwi.detr.gov.uk/soslist.

Chapter 14

1. ENVIRONMENT ACT, 1995.
2. WATER INDUSTRY ACT, 1999.
3. LAMBERT A.O., BROWN T.G., TAKIZAWA M. and WEIMAR D. A review of performance indicators for real losses from water supply systems. *J Water SRT—Aqua*, 1999, **48**(6), 227–237.
4. WRC *Managing leakage*. 1994, ISBN 1-898920-06-0.
5. *Water Supply (Water Fittings) Regulations Statutory Instrument*. No. 1148, 1999, HMSO, London.
6. OFWAT *Leakage and the efficient use of water 1999–2000 report.* Birmingham, 2000, ISBN 1-874234-69-8.

Index